An Award Winning Math Book
For Excellence
Florida Publishers Group
and
National Association of
Independent Publishers
Book of the Year

WINNING AT MATH

WINNING AT MATH

Your Guide To Learning Mathematics Through Successful Study Skills

by Paul D. Nolting, Ph.D.

ACADEMIC SUCCESS PRESS, INC.
POMPANO BEACH, FLORIDA

WINNING AT MATH
BY PAUL D. NOLTING

Copyright ©1991 by Paul D. Nolting, Ph.D.

Published by Academic Success Press, Inc.
Produced by Wright & Ratzlaff Associates
Cover Design by Ann Boughton
Book Design by DeAun Downs
Illustrations by James B. King
Printed in the United States of America
New, Expanded and Revised Second Edition, 1991.

Library of Congress Cataloging-in-Publication Data

Nolting, Paul D., 1951-
    Winning at math : your guide to learning mathematics through successful study skills / by Paul D. Nolting. -- New, expanded, and rev. 2nd ed.
        p.   cm.
    Includes Index.
    ISBN: 0-940287-19-6 (pbk.) : $12.95
    1. Mathematics -- Study and teaching.   I. Title.
QA11.N583   1991
510'.71 -- dc20   91-19306
                                                                CIP

CONTENTS

PREFACE

Mathematics is one of the most difficult subjects in college. Every student must pass mathematics to graduate. In fact, some colleges require students to take up to four mathematics courses to obtain an Associate of Arts degree. Too many students are intimidated by mathematics because they have a poor high school mathematics background or are returning to school after many years.

What kind of assistance do most students want? Students want concrete tips and procedures to help them improve their grades in math. *Winning at Math* provides those tips and procedures and has been proven effective in improving a student's ability to learn mathematics and take tests.

The study of anxiety reduction and test-taking procedures provided in *Winning at Math* are based on Learning Specialist Dr. Paul D. Nolting's ten years of experience working with students who have had difficulty with mathematics. These techniques are statistically effective with students who have previously failed mathematics. These techniques can definitely work for you!

xiii

INTRODUCTION

Studying and learning mathematics is different from other courses. For this reason many students make A's and B's in all other courses but have difficulty passing mathematics. By using the suggested study procedures in *Winning at Math* you will be able to improve your mathematics comprehension and make better grades. If you have previously failed math, *Winning at Math* can greatly increase your chances of passing.

You can win with *Winning at Math* by studying it on your own, using it as a part of a traditional course or as an independent study program through a mathematics lab. Combined with the cassette tape *How to Reduce Test Anxiety,* * *Winning at Math* becomes even more effective. This combination provides for a more complete mathematics study program.

To discover your mathematics learning strengths and weaknesses, complete the Math Pre-Test-Study Skills Evaluation. The self-scoring section will suggest the exact pages in *Winning at Math* you should first read. For a more complete diagnosis of your math learning problems and a personalized printed prescription for success use the *Winning At Math: Study Skills Computer Evaluation Software.*

After reading each chapter, complete the reading assignments and questions located on the last page of each chapter. By reading *Winning At Math* and completing the assignments you can improve your ability to learn mathematics dramatically.

HOW TO REDUCE TEST ANXIETY ($15.95 postpaid) and *Winning At Math: Study Skills Computer Evaluation Software* ($52.95 postpaid) are available from the Academic Success Press, Inc., P. O. Box 2567, Pompano Beach, FL 33072, tel: (305) 785-2034.

MATH PRE-TEST STUDY
SKILLS EVALUATION

Read each of the items below. Choose the statement in each group which is true of you. Indicate what you *actually do* rather than what you *should do* by circling a, b, or c. BE HONEST.

1. I:
   a. seldom study math every school day.
   b. often study math every school day.
   c. almost always study math every school day.

2. When I register for a math course, I:
   a. seldom select the best math teacher.
   b. often select the best math teacher.
   c. almost always select the best math teacher.

3. I:
   a. seldom become anxious and forget important concepts during a math test.
   b. often become anxious and forget important concepts during a math test.
   c. almost always become anxious and forget important concepts during a math test.

4. I:
   a. seldom study math at least 8 to 12 hours a week.
   b. often study math at least 8 to 12 hours a week.
   c. almost always study math at least 8 to 12 hours a week.

5. Each week, I:
   a. seldom plan the best time to study math.
   b. often plan the best time to study math.
   c. almost always plan the best time to study math.

6. I:
   a. seldom use an abbreviation system when taking notes.
   b. often use an abbreviation system when taking notes.
   c. almost always use an abbreviation system when taking notes.

7. When I take math notes, I:
   a. seldom copy all the steps to a problem.
   b. often copy all the steps to a problem.
   c. almost always copy all the steps to a problem.

8. When I become confused in math class, I:
   a. seldom stop taking notes.
   b. often stop taking notes.
   c. almost always stop taking notes.

9. I:
   a. seldom fail to ask questions in math class.
   b. often fail to ask questions in math class.
   c. almost always fail to ask questions in math class.

10. I:
    a. seldom stop reading the math textbook when I get stuck.
    b. often stop reading the math textbook when I get stuck.
    c. almost always stop reading the math textbook when I get stuck.

11. When I have difficulty understanding the math topic, I:
    a. seldom go to the instructor or tutor.
    b. often go to the instructor or tutor.
    c. almost always go to the instructor or tutor.

12. I:
    a. seldom review class notes or read the textbook assignment before doing my homework.
    b. often review class notes or read the textbook assignment before doing my homework.
    c. almost always review class notes or read the textbook assignment before doing my homework.

13. I:
    a. seldom fall behind in completing math homework assignments.
    b. often fall behind in completing math homework assignments.
    c. almost always fall behind in completing math homework assignments.

14. After reading the math textbook, I:
    a. seldom mentally review what I have read.
    b. often mentally review what I have read.
    c. almost always mentally review what I have read.

15. There:
    a. seldom are distractions that bother me when I study.
    b. often are distractions that bother me when I study.
    c. almost always are distractions that bother me when I study.

16. I:
    a. seldom do most of my studying the night before the test.
    b. often do most of my studying the night before the test.
    c. almost always do most of my studying the night before the test.

17. I:
    a. seldom develop memory techniques to remember math concepts.
    b. often develop memory techniques to remember math concepts.
    c. almost always develop memory techniques to remember math concepts.

18. When taking a math test, I:
    a. seldom start on the first problem and work the remaining problems in their numbered order.
    b. often start on the first problem and work the remaining problems in their numbered order.
    c. almost always start on the first problem and work the remaining problems in their numbered order.

19. Even when time permits, I:
    a. seldom check over my test answers.
    b. often check over my test answers.
    c. almost always check over my test answers.

20. When my math test is returned, I:
    a. seldom analyze the test errors.
    b. often analyze the test errors.
    c. almost always analyze the test errors.

MATH POST-TEST SCORING
FOR STUDY SKILLS

Put the correct amount of points for each item in Section A and Section B to obtain your score. The order of the items are different for Section A and Section B. To the right of each item's score are page numbers where the information on this topic is discussed.

SECTION A POINT VALUE EACH STATEMENT

Items	Answer A (1 point)	Answer B (3 points)	Answer C (5 points)	*Winning At Math* Reference
1.	————	————	————	pp. 21-24 101-102
2.	————	————	————	pp. 37-38
4.	————	————	————	pp. 69-74
5.	————	————	————	pp. 75-77
6.	————	————	————	pp. 84-85
7.	————	————	————	pp. 89-90
11.	————	————	————	pp. 98-99
12.	————	————	————	pp. 100-101
14.	————	————	————	pp. 110
17.	————	————	————	pp. 127-131
19.	————	————	————	pp. 140
20.	————	————	————	pp. 140-145

TOTAL ———— + ———— + ———— = ————————
SECTION A

SECTION B POINT VALUE FOR EACH STATEMENT

Items	Answer A (5 points)	Answer B (3 points)	Answer C (1 point)	*Winning At Math* Reference
3.	————	————	————	pp. 59-65
8.	————	————	————	pp. 87-88
9.	————	————	————	pp. 89
10.	————	————	————	pp. 97-99
13.	————	————	————	pp. 101-102
15.	————	————	————	pp. 117-119
16.	————	————	————	pp. 120-121
18.	————	————	————	pp. 138-139
TOTAL	———— +	———— +	———— =	————————— SECTION B

———— + ———— = —————————

SECTION A   SECTION B   GRAND TOTAL

A score of 70 or below means you have poor math study skills.

A score between 70 and 90 means that you have good study skills, but you can improve.

A score above 90 means that you have excellent math study skills.

After completing the Math Post-Test Study Skills Evaluation, compare your score to your Math Pre-Test Study Skills Evaluation score. A great improvement in the Post-Test score indicates an improvement in math study skills. If you did not score a 100 on the Post-Test Study Skills Evaluation, then review all the questions that you scored one or three points. Review all the pages in *Winning At Math* that are indicated in this survey to improve your math study skills. You can also use the Strategy Cards for Higher Grades for an excellent review of your math study skills.

Chapter 1
Understanding Special Study
Requirements for Mathematics

Mathematics courses are considered to be totally different from other college courses and require different study procedures. Passing most of your other college courses requires only that you read and understand the subject material. However, to pass mathematics, an extra step is required: applying the material by doing the problems.

EXAMPLE: Political science courses require reading the textbook and understanding the material. But your instructor isn't going to make you run for political office to apply knowledge you obtained.

In mathematics you must understand the material, comprehend the material and apply the material. Applying mathematics is the hardest task.

Linear Learning Pattern

Another characteristic of mathematics is its linear learning pattern. *Linear learning pattern means that the material learned on one day is used the next day and the next day, and so forth.*

If you fail to understand the classroom material the first week, you may never catch up. Linear learning effects studying for tests in mathematics as well. If you study Chapter One and understand it, study Chapter Two and understand it, and study Chapter Three and *do not understand it*, then when you have a test on Chapter Four, you're not going to understand it either.

In a history class, if you study for Chapter One and Chapter Two, and do not understand Chapter Three, and end up studying and having a test on Chapter Four, you could pass. Understanding Chapter Four in history is not totally based on comprehending Chapter Three.

To succeed in mathematics each previous chapter has to be understood before continuing to the next chapter.
When students get behind in mathematics it is difficult to catch up. Mathematics learning is a building process. All building blocks must be included to win at math. Mathematics learning builds up geometrically and compounds itself. Math is not a subject in which you can forget the material after a test. REMEMBER: To learn the new math material for the test on Chapter Five, first you must go back and learn the material in Chapter Four. This means you will have to go back and learn Chapter Four while learning Chapter Five. The best of us can fall behind under these circumstances. However, if you do not understand the material in Chapter Four, you will not understand the material in Chapter Five either and will fail the test on Chapter Five.

Math as a Foreign Language

Another way to understand studying for mathematics is to consider it a foreign language. Looking at mathematics as a foreign language can improve your study procedures. In the case of a foreign language, if you do not practice it, what happens? You forget it. If you do not practice mathematics what happens? You are likely to forget it too. Students who excel in a foreign language study and practice it at least every other day. The same study habits apply to mathematics, because it is considered a foreign language.

Like a foreign language, mathematics has unfamiliar vocabulary words or terms to be put in sentences called expressions or equations. Understanding and solving a mathematics equation is the same as speaking and understanding a foreign language. Mathematics sentences have words in them, such as equal (=), less (−), and unknown (a).

Learning how to speak mathematics as a language is the key to success. Currently most universities consider computer and statistics (a form of mathematics) courses as foreign languages. Universities have now gone as far as to make mathematics a foreign language.

Mathematics is not a popular topic. You do not hear Dan

"YOU DO NOT HEAR DAN RATHER ON TV TALKING IN MATHEMATICS FORMULAS"

Rather on TV talking in mathematics formulas. He talks about major events in countries like Korea to which we can relate politically, geographically, and historically. Through TV — the greatest of learning tools — we learn English, humanities, speech, social studies, and natural sciences but not mathematics. Mathematics concepts are not constantly reinforced like

English or other subject areas in our everyday lives. Mathematics has to be learned independently. Therefore, it requires more study time.

High School vs. College Math

Mathematics as a college level course is almost two to three times as difficult as high school mathematics. In college, the Fall and Spring math class time has been cut to three hours a week. High school math gives you five hours a week. Furthermore, college courses are taught twice as fast as high school courses; what is learned in one year in high school is learned in one semester (four months) in college. This enhances mathematics study problems for the college student; you are receiving less instructional time and proceeding twice as fast. The responsibility for learning mathematics has now shifted from the school to the student, and most of your learning will have to occur outside the college classroom.

Summer vs. Fall or Spring Semesters

Mathematics courses taught in summer semesters are more difficult than Fall or Spring semester courses. Students in a six week summer session must learn math two and a half times as fast as regular semester students. Though you receive the same amount of instructional classroom time, there's less time to understand the material between class sessions. Summer semester classes are usually two hours a day and four days a week. If you don't understand the lecture on Monday, then you only have Monday night to learn the material before progressing to more difficult material on Tuesday. Since mathematics is a linear learning experience where every building block must be understood, you can fall behind quickly and never catch up. In fact, some students become *lost* during the first half of a math lecture and never understand the rest of the lecture. This is called *kamikaze* math, since most students don't survive the course. *If you have to take a summer mathematics course, take a ten or twelve week session so that you have more time to process the material between classes.*

Course Grading System

The course grading system for mathematics is different in college than in high school. While in high school, if you make a "D" or borderline "D/F" the teacher more than likely will give you a "D" and you may go on to the next course. However, in some college mathematics courses, students cannot make a "D", or, if a "D" is made, the course will not count towards graduation. Also, college instructors are more likely to give an "N" (no grade), "W" (withdraw from class), or "F" for barely knowing the material, because you will be unable to pass the next course.

Most colleges require students to pass two college level algebra courses to graduate. In high school you may graduate by passing one to three arithmetic courses. In college you might have to take four mathematics courses and make "C's" in all of them to graduate. *The first two high school mathematics courses will be preparation for the two college level algebra courses. Therefore you must dramatically increase the quality and quantity of your mathematics study skills to pass more mathematics courses with higher grades.*

Your First Math Test

Making a high grade on the first major math test is more important than making a high grade on the first major test in other college subjects. The first major math test taken is the easiest and most often least prepared for. Students feel that the first major math test is mainly review and they can make a "B" or "C" without much study. These students are overlooking an excellent opportunity to make an "A" on the easiest major math test of the semester which counts the same as the more difficult remaining major math tests. These students, at the end of the semester, sometimes do not pass the math course or do not make an "A" because their first major test grade was not high enough to pull up a low test score on one of the remaining major tests.

Studying hard for the first major math test and obtaining an "A" has several advantages.

★ A high score on this math test *can compensate for a low*

score on a more difficult fourth or fifth math test — and major tests count the same.

★ Knowing you have learned the basic math skills required to pass the course. *This means you will not have to spend time relearning the misunderstood material covered on the first major test while learning new material for the next test.*

★ *Improving motivation for higher test scores.* Improved motivation can cause you to increase your math study time allowing you to master the material.

★ *Improving confidence for higher test scores.* With more confidence you are more likely to work harder on the difficult math homework assignments which will increase your chances of doing well in the course.

College math instructors treat students differently than high school mathematics instructors. High school mathematics teachers warn you about your grades. Some college instructors may ask you, "How are you doing in the course?" Do not expect them to say, "You have been making "D"'s and "F"'s on your tests and you need to come to see me." That would be a rare response. *You must take responsibility and make an appointment to seek help from your instructor.*

Sometimes, due to the increase in the number of college math courses, there are more adjunct math faculty than fulltime math faculty. This problem can restrict student and instructor interaction. Fulltime faculty have regular office hours and are required to help students a certain number of hours per week in their office or math lab. However, adjunct faculty are only required to teach their mathematics courses; they don't have to meet students after class even though some adjunct faculty will provide this service. Since mathematics students usually need more instructor assistance after class than other students, having an adjunct math faculty member could require you to find another source of course help. *Try to select a fulltime math faculty member as your instructor.*

Finding a Study Buddy

Getting a study buddy is suggested. You can't always depend

on having the instructor available for help. A study buddy is someone in your class to call when you have difficulty working mathematics problems. *A study buddy can improve your study time.*

SUMMARY

★ Learning mathematics requires different skills than learning other courses. Learning mathematics is like learning a foreign language.

★ Passing most courses requires only reading and understanding the subject material. However, when taking a mathematics test, you not only have to understand and comprehend the material, you have to prove this to the instructor by working the problems.

★ In most other tests you can just guess at the answers. You can't guess at the answers in mathematics tests. The answers must be precise and exact.

★ Mathematics courses are even more difficult because a grade of "C" or better is usually required to take the next course, or in some cases, just to pass it.

ASSIGNMENT FOR CHAPTER 1

1. Why is math considered to have a linear learning pattern?

2. How is math similar to a foreign language?

3. How are high school and college math courses different?

4. How are Summer and Fall math courses different?

5. Where can you obtain additional help for your math course?

6. Who can you find to be your study buddy?

7. What are your math instructor's policies about missing classes and tests?

NOTES

Chapter 2

Identifying Your Positive and Negative Mathematics Achievement Characteristics

Mathematics achievement characteristics are characteristics which effect students' grades, such as previous math knowledge, level of test anxiety, study habits, study attitudes, motivation, and test-taking skills. Before we start identifying your mathematics achievement characteristics you need to understand what contributes to mathematics academic success. Dr. Benjamin Bloom, a famous researcher in the field of educational learning, discovered that IQ and cognitive entry skills account for 50% of your course grade (See **Figure 1**). Quality of instruction represents 25% of your course grade, while affective student characteristics account for the remaining 25% of your grade.

IQ is relatively stable but may be considered, for instance, as how fast the student can relearn old algebra or learn new algebra concepts.

Cognitive entry skills refer to how much math a student knows before entering the course.

Affective characteristics are characteristics students possess that effect their course grades — excluding cognitive entry skills. Some of these characteristics are anxiety, study habits, study attitudes, self-concept, motivation and test-taking skills.

Quality of instruction is concerned with the effectiveness of math instructors when presenting material to students in the classroom and math lab. This effectiveness depends on the course textbook, teaching style, extra teaching aids (videos, audio cassettes), and other assistance.

Assessing Your Math Cognitive Entry Skills

Cognitive entry skills can cause low grades if a student is placed in a math course that requires a more extensive mathematics background or quicker learning skills than the

FIGURE 1
VARIABLES CONTRIBUTING TO STUDENT
ACADEMIC ACHIEVEMENT

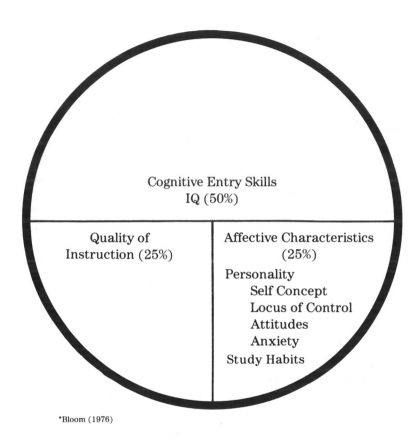

*Bloom (1976)

student has.

Required cognitive skills to enroll in a mathematics course tend to be measured by placement test (ACT, SAT) or the grade received in the prerequisite math course. These students try to encourage their instructors to move them to a higher level math course believing they have received an inaccurate low placement score. Other students try to avoid non-credit math courses while other students don't want to repeat a course that they've previously failed.

Some older students imagine their math skills are just as good as they were five to ten years ago when they were enrolled in their last high school or college math course. If they haven't been practicing their math skills, they are just fooling themselves. Other students believe they don't need the math skills obtained in a prerequisite math course to pass the next course, which is also incorrect. *Research indicates that students who were placed correctly in their prerequisite math course and who failed it will not pass the next math course.*

I have conducted research on thousands of students who have either convinced their instructors to place them in higher level math courses or have placed themselves in higher level math courses. The results? These students failed their math courses many times before realizing they didn't possess the prerequisite cognitive entry skills needed to pass the course.

To be successful in a math course you must have the appropriate cognitive entry skills. Thinking you may have difficulty passing a higher level math course means you probably don't have an adequate math background. Even if you do pass the math course with a "D" or "C", research indicates you will most likely fail the next higher level math course.

It is better to be conservative and pass a lower level math course with an "A" or "B" instead of making a "C" or "D" in a higher level math course and failing the next math course.

This is evident by the many students who repeat a higher level math course up to five times before repeating the lower level math course that was barely passed. After repeating the lower level math course with an "A" or "B", these students passed their higher level math course easily.

Improving Your Cognitive Entry Skills

There are several ways to improve your cognitive mathematics entry skills. *Review your previous math course material before attending your present math course.*

Instructors always think you finished the previous math course last week; they don't wait for you to catch up on current material. It doesn't matter if your previous math course was a month ago or five years ago, instructors expect you to know the previous course material.

To make a high grade on your first math test, you must possess the same math skills as when you finished your previous math course. You can obtain these skills by reviewing old math tests or working problems in the previous textbook's chapter reviews. If your last math course was over a year ago, a tutor may be required to elevate your skills to a competent level.

Another way in which to obtain appropriate mathematics cognitive entry skills is to take each math course "back to back." It is better to take math courses every semester — even if you don't like math — so that you can maintain the linear learning sequence. I have known students who have made "B's" or "C's" in a math class, then waited six months to a year to take the next math course; inevitably they fail. These students didn't complete any preparatory math work before the math course and were lost after the second chapter. This is similar to having one semester of Spanish, not speaking it for a year, then visiting Spain, and not understanding what is being said. The only exception to taking math courses "back to back" is taking a six week kamikaze math course, which should be avoided.

If you are one of the unfortunate many who are currently failing a math course, you need to ask yourself, "Am I currently learning any math or just becoming more confused?" If you are learning some math, stay in the course. If you are getting more confused, then withdraw from the course. Improve your cognitive entry skills for the next semester.

EXAMPLE: You have withdrawn from a math course after midterm due to low grades. Instead of closing the math book until next semester, attend a math lab or seek a tutor and learn Chapters One, Two, and Three to perfection. You will enter the same math course next semester with excellent cognitive entry skills. In fact, you can make an "A" on the first test and complete the course with a high grade. Does this sound far-fetched? It may, but I know hundreds of students who have used this learning procedure and passed their math course instead of failing the same math course again and again.

Hiring A Private Tutor

One last way to improve your cognitive entry skills is to

employ a private tutor. If you have a history of not doing well in math courses, then you need to start tutorial sessions the *same* week class begins. This will give the tutor a better chance of helping you regain those old math skills.

You will need to work hard to relearn those old math skills, while continuing to learn the new material. If you wait four to five weeks to obtain a tutor, it will probably be too late to catch up and do well or even pass the course. *Tutorial sessions work best in improving cognitive entry skills when the sessions begin during the first two weeks of a math course.*

Quality of Instruction

Quality of instruction can effect your math grade in several ways. Classroom atmosphere, instructor's teaching style, lab instruction, textbook content and format can all effect your ability to learn in the math classroom. Interestingly enough, the most important variable is the compatibility of an instructor's teaching style with your learning style. Noncompatibility can be best solved by finding another instructor who better matches your learning style.

Finding the Best Instructor

Finding an instructor who matches your learning style can be a difficult task. Your learning style is important; your learning style is how you acquire information best. For example, some students learn better through hearing the information over and over again, instead of reading the information carefully.

Most students are placed in their first math course by an academic advisor. Usually academic advisors know who are the best and worst math instructors. However, advisors can be reluctant to discuss teacher qualities. Unfortunately, students may want the counselor to devise a course schedule based on the student's time limits instead of teacher selection.

To learn who are the best math instructors, ask the academic advisor which math instructor's classes fill up first. This doesn't place the academic advisor in the position of making a value judgment; neither does it guarantee the best instructor. But it will increase the odds in your favor.

Another manner in which to acquire a good math instructor is to ask your friends about their current and previous math instructors. However, if a fellow student says an instructor is excellent, make sure your learning style matches your friend's learning style. Ask your friend, "Exacly what makes the instructor so good?" Then, compare the answer to how you learn best. If you have a different learning style than your friend, look for another instructor.

To obtain the most from an instructor, discover your best learning style and match it to the instructor's teaching style. Most learning centers or student personnel offices will have counselors who can measure and explain your learning style. Then, interview or observe the instructor while the instructor is teaching. This process is time consuming, but it is well worth the effort!

Once you have found your best instructor, don't change. Remain with the same instructor for every math class. While this may be an inconvenience, it will pay off with higher grades.

Affective Characteristics

By working to improve your affective characteristics, direct benefit can be seen in higher scores in mathematics. The major affective characteristics needing improvement tend to be study habits, anxiety level, and control over mathematics. However, most students are not taught how to improve their affective characteristics until they reach college. You must learn *now* to be more effective in winning at math by improving your affective characteristics!

Measuring Your Math Strengths and Weaknesses

Students have positive and negative mathemathics achievement characteristics that effect their course grades. FOR EXAMPLE: Development of good study habits is a positive mathematics achievement characteristic, while high test anxiety is a negative mathematics achievement characteristic.

To learn your academic strengths and weaknesses, you may

"MEASURING YOUR MATH STRENGTHS AND WEAKNESSES"

have the opportunity to take some surveys and tests such as the Mathematics Anxiety Rating Scale (MARS), Survey of Study Habits and Attitudes (SSHA), Nowicki-Strickland Locus of Control (NSLC), Nelson Denny Reading Test (NDRT), and the mathematics portion of the Wide Range Achievement Test (WRAT).

Students who don't have the opportunity to take these tests and surveys need to locate alternate ways in which to measure their mathematics achievement characteristics. Consult your local high school counselor, college counselor, learning specialist or tutor center specialist for assistance in having your math achievement characteristics measured.

Different types of tests and surveys may be used to measure the same concepts:

★ MARS is a survey for assessing college mathematics anxiety and will produce a score indicating how much anxiety you have, compared to other college students.

★ SSHA is a survey measuring your study habits and attitudes compared to other college students who make "A's" in their courses.

★ NSLC is an opinion survey which estimates how much control you believe you have over life events. Internal students take resposibility for their grades and try to improve their learning skills. External students believe they have little control over their grades and blame the school or others for their poor achievement.

★ NDRT is a test which measures college vocabulary and reading comprehension, compares your scores to other college students, and gives a grade level equivalent.

★ WRAT (the mathematics portion) measures arithmetic and algebra skills and gives your score as a grade level equivalent.

These surveys and tests give an appropriate measurement of the amount of cognitive and affective skills you possess.

Once you have completed all the surveys and tests, enter the scores on "Your Student Profile Sheet" (See **Figure 2**). The Student Profile Sheet will reveal your academic strengths and weaknesses. Your course instructor will explain the meaning of the test and survey scores as they relate to improving your grades in mathematics by suggesting the academic areas where you need improvement.

If you cannot complete the suggested tests and surveys, ask your instructor or counselor about taking the *Winning At Math: Study Skills Computer Evaluation Software*. This computerized math study skills evaluation will diagnose your speccific math-learning problems. A personalized printed prescription for success will be generated with brief explanations of your math-learning problems. Read each explanation to understand your learning problems. Follow the suggestions on the print-out in the areas you receive one point. These one-point areas indicate your worst study skills and should be improved immediately.

FIGURE 2

YOUR STUDENT PROFILE SHEET

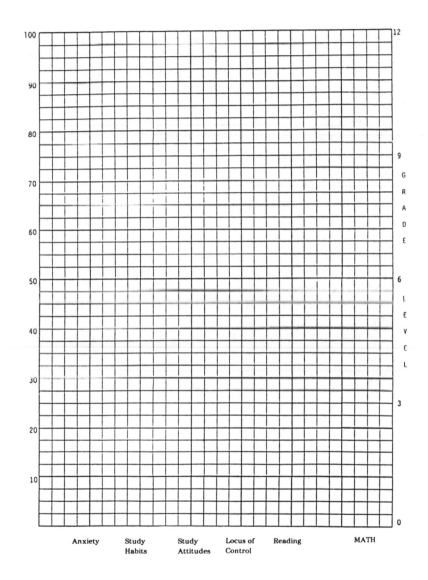

If you can't obtain any tests or surveys to measure your mathematics achievement characteristics, then *you must assume that you need to improve all the learning and relaxation skills* discussed in *Winning at Math*.

To better understand the Student Profile Sheet, look at **Figure 3** which represents the profile of a 30-year-old married student who works part time and has a family. The student's profile graph contains mathematics anxiety, study habits, study attitudes, locus of control, and reading level scores based on college percentile norms. The mathematics test measured her grade level equivalent. A high score on the MARS anxiety scale indicates high test anxiety. High scores on the other scales indicate success in those areas.

According to this student's profile, she has extremely high mathematics anxiety, poor study habits, good study attitudes, internal locus of control, average reading level, and about a 9th grade mathematics level. Therefore, she believes that she can make a good grade in math and has a good attitude about school. This will help improve her study skills and decrease mathematics anxiety. In addition, her reading level and mathematics level are about average and will not present major learning blocks towards improving her math grades.

RESULTS: This student learned how to decrease her mathematics anxiety and improve her study skills while attending my study skills class. She had failed her algebra course twice before taking my class. After completing my class, she took the course again and passed with a "B"!

FIGURE 3

STUDENT PROFILE SHEET

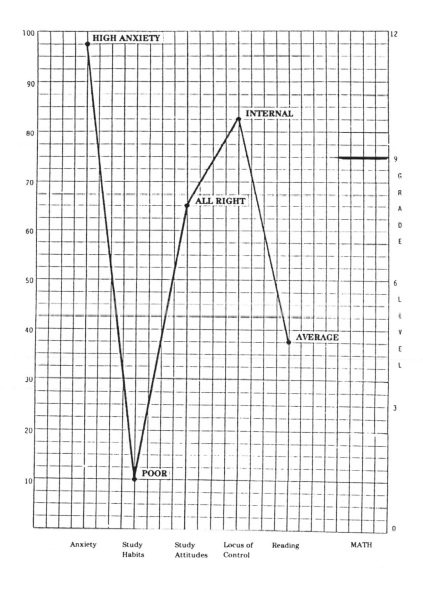

Figure 4 represents a profile of a student who has moderate test anxiety, average study habits and excellent study attitude. The student's locus of control and reading ability are average, while she has a ninth grade math level. Her problem with passing math is similiar to the student's profile in **Figure 3**, though not as drastic. However, the student in **Figure 4** still has to decrease her test anxiety, improve her study habits, and become more internal in order to be successful in math.

FIGURE 4

STUDENT PROFILE SHEET

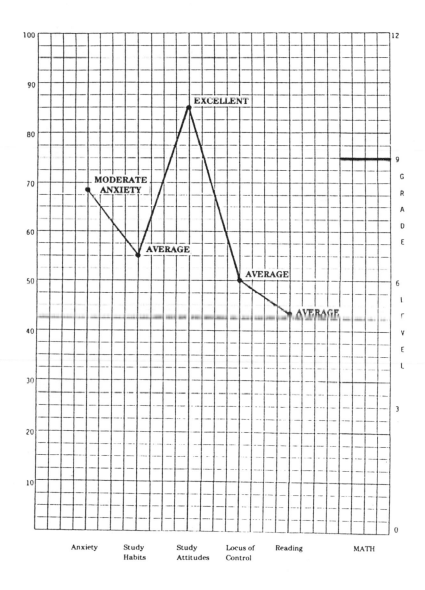

Figure 5 represents a student who had a long history of failing math. The only two positive characteristics of this student are his low test anxiety and average reading level. He has poor study habits, poor study attitude, and is external in his locus of control — all probably due to failing math so many times. His reading and math skills are average. This external student must believe he can pass math through improving his study habits and attitudes. He also needs support from his teacher and counselor to become more internal and pass math.

FIGURE 5

STUDENT PROFILE SHEET

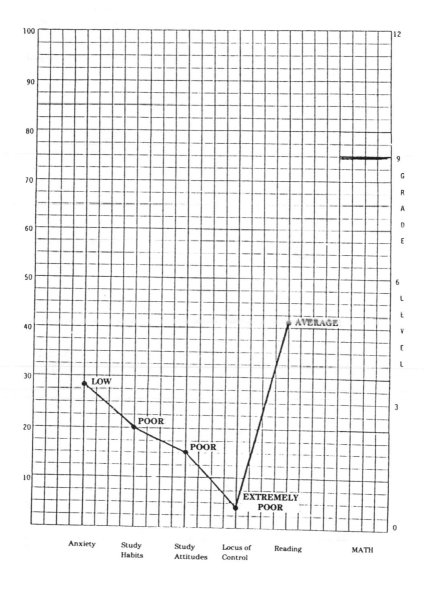

Figure 6 represents a student who has low anxiety, excellent study attitude, excellent reading level and an internal locus of control. His math level is average. It appears that his poor study habits are the major block to becoming successful in math.

His poor study habits consist of procrastination in studying math and not having efficient study skills. He is motivated to learn math and only needs to learn how not to procrastinate and how to improve his efficiency in studying.

FIGURE 6

STUDENT PROFILE SHEET

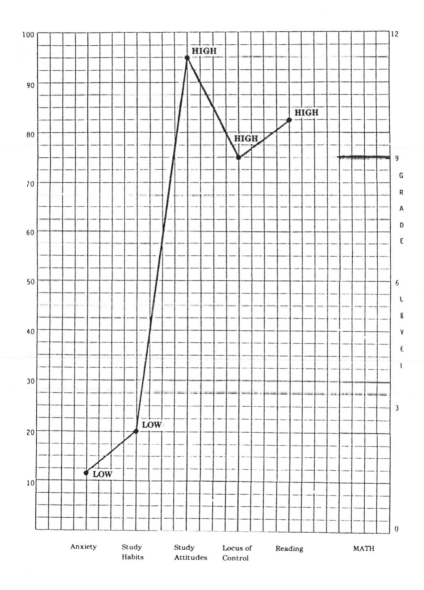

Figure 7 is from a middle age male student who has average study habits and locus of control. He has an excellent study attitude but has high test anxiety, a poor reading level, and a below average math level. High test anxiety and a low reading level are his major blocks to learning math.

RESULTS: This student was referred to the reading lab and math lab to improve his basic learning skills. He also decreased his test anxiety and improved his study habits.

FIGURE 7

STUDENT PROFILE SHEET

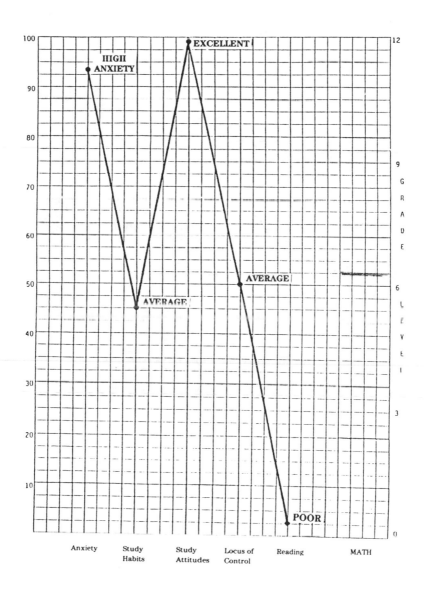

From these student profiles it is evident that each student has different reasons for being unsuccessful at math. Their problems usually occur in one of the following areas:

(1) high anxiety

(2) poor study habits

(3) poor study attitude, or

(4) external locus of control.

SUMMARY

Mathematics success is primarily based on improving the characteristics that affect your ability to learn math — your affective learning characteristics. The major affective learning characteristics are study habits, anxiety, and control over mathematics.

There are various ways to improve your cognitive entry skills and quality of instruction. Once placed in the appropriate mathematics course, success is based on your ability to learn math. Learning which affective learning characteristics you need to improve may be accomplished by completing your Student Profile Sheet.

ASSIGNMENT FOR CHAPTER 2

1. If possible, take the MARS, SSHA, NDLC, NDRT, WRAT (mathematics only) surveys or tests.

2. Review your Student Profile Sheet and Math Study Skills Evaluation with the course instructor.

3. How are your math cognitive entry skills?

4. How can you improve your math cognitive entry skills?

5. How can you find the best math instructor for you?

6. Describe your positive and negative math achievement characteristics from your student profile sheet or Math Study Skills Evaluation.

7. What steps are you taking to improve your negative math achievement characteristics?

8. Who is your study buddy?

NOTES

NOTES

Chapter 3
Learning How to Reduce
Mathematics Test Anxiety

Test anxiety has existed for as long as tests have been used to evaluate grades. Over the last 45 years, test anxiety has been studied. Pioneering studies, in general, indicate that test anxiety leads to low test scores.

At the University of South Florida, Dr. Charles Spielberger investigated the relationship between test anxiety and intellectual ability. The study suggested that anxiety coupled with high ability can improve academic performance, but anxiety coupled with low or average ability can interfere with academic performance:

Anxiety + High Ability = Improvement

Anxiety + Low or Average Ability = No Improvement

Average ability students with low test anxiety had better performance and higher grades than did average ability students with high test anxiety.

Text anxiety is a learned response and is a sub-category of general stress. General stress is considered to be strained exertion which can lead to physical and psychological problems. For a better understanding of general stress, read Reference B.

Defining Test Anxiety

There are several definitions of test anxiety. One definition states, "Test anxiety is a conditioned emotional habit to either a single terrifying experience, recurring experience of high anxiety, or a continuous condition of anxiety." A person isn't born with test anxiety. An environmental situation brings about that anxiety. *Test anxiety is a learned response that can be unlearned.*

Another definition of test anxiety is related to the educational system. This definition suggests test anxiety is the anticipation of some realistic or non-realistic situational threat. The test can be a research paper, oral report, work at the blackboard, multiple choice or essay.

Learning the Causes of Test Anxiety

The causes of test anxiety can be different for each student but can be explained by four basic concepts:

1. Test anxiety can be a learned behavior resulting from the expectations of parents, teachers or other significant people in the student's life.

2. Test anxiety can be caused by the association between grades and a student's personal worth.

3. Test anxiety develops from fear of alienating parents, family, or friends due to poor grades.

4. Test anxiety can stem from a feeling of lack of control and an inability to change one's life situation.

Mathematics test anxiety is a new concept in education. *Ms. Magazine* (1976) published "Math Anxiety: Why is a Smart Girl Like You Counting on Your Fingers?" and coined the phrase math anxiety to the general public. During this period other educators began using the terms mathophobia or mathemaphobia as a possible cause for children's unwillingness to learn mathematics. Additional studies on the graduate level discovered mathematics anxiety was common among adults. Educators did not focus on mathematics anxiety as a psychological state, but as a skill deficiency, until the *Ms. Magazine* article appeared.

There are several definitions of mathematics anxiety:

* Mathematics anxiety is generally considered the extreme of a very negative attitude toward mathematics. There is a high relationship between low mathematics confidence and high mathematics test anxiety; this could be the same concept.

* Mathematics anxiety is also defined as the feeling of tension and anxiety that interferes with the manipulation of numbers and the solving of mathematical problems in academic test situations.

* Mathematics anxiety can also be defined as a state of panic, helplessness, paralysis, and mental disorganization

that occurs in some students when required to solve mathematics problems. This discomfort varies in intensity and is the outcome of numerous previous situations.

Mathematics anxiety can be divided into mathematics text anxiety and numerical anxiety. Mathematics text anxiety involves anticipation, completion, and feedback while numerical anxiety refers to everyday situations requiring some form of number manipulation.

It has been shown that mathematics anxiety exists among many students who usually do not suffer from other tensions. *Counselors reported that one-third of the students responding to behavior therapy programs offered through counseling centers had problems with mathematics anxiety.*

Educators know that mathematics anxiety occurs frequently among college students and is more prevalent in women than men, and also frequently occurs in students having poor high school mathematics backgrounds; these students have the greatest amount of anxiety. *Approximately half the students in college prep mathematics courses, designed for students with inadequate high school mathematics background or low placement scores, could be considered to have mathematics anxiety.*

Educators investigating anxiety and mathematics achievement have indicated that anxiety contributes to poor grades in mathematics, especially for women. Educators also indicate that just reducing mathematics test anxiety does not guarantee higher mathematics grades. Students must also have good mathematics study skills to first learn the mathematics material to be used later on tests.

Reducing Test Anxiety

To reduce mathematics test anxiety, an understanding of the relaxation response is required. The relaxation response is any technique or procedure that enables you to become relaxed and will take the place of an anxiety response.

The two basic types of test anxiety controlled by the relaxation response are somatic and cognitive anxiety. Somatic anxiety occurs in your body. Signs of somatic anxiety are upset stomach, sweaty palms, pain in the neck, or general feelings of nervousness. These feelings of nervousness prevent you from

"REDUCING TEST ANXIETY"

totally concentrating on the test. Cognitive anxiety is negative self-talk that distracts you from concentrating on the test.

FOR EXAMPLE, during a test you keep telling yourself. "I can not do it, I can not do the problems, and I am going to fail this test."

There are short-and long-term relaxation response techniques that help control mathematics test anxiety. Effective short-term techniques include the tensing and differential relaxation method, and the palming method.

The tensing and differential relaxation method helps you relax by tensing and relaxing your muscles all at once. Follow these procedures while you are sitting at your desk before taking a test:

1. Put your feet flat on the floor.

2. With your hands, grab under the chair.

3. Push down with your feet and pull up on your chair at the same time for about five seconds.

4. Relax for five to ten seconds.

5. Repeat two-three times.

6. Relax all your muscles except the ones that are actually used to take the test.

The palming method is a visualization procedure used to reduce test anxiety. While you are at your desk before or during a test, follow these procedures:

1. Close your eyes and cover them using the center of the palms of your hands.

2. Prevent your hands from touching your eyes by resting the lower parts of the palms of your cheekbones and placing your fingers on your forehead. The eyeballs must not be touched, rubbed or handled in any way.

3. Think of some real or imaginary relaxing scene. Mentally visualize this scene. Picture the scene as if you were actually there, looking through your own eyes.

4. Visualize this relaxing scene for one-two minutes. Practice visualizing this scene several days before taking a test and the effectiveness of this relaxation procedure will improve.

Side one of the audio cassette, *How to Reduce Test Anxiety* further explains test anxiety and discusses these and other short-term relaxation response techniques. These short-term relaxation techniques can be learned quickly but are not as successful as the long-term relaxation technique. Short-term techniques are intended to be used while learning the long-term technique.

The cue-controlled relaxation response technique is the best long-term relaxation technique. It is presented on "Side Two" of the audio cassette, *How To Reduce Test Anxiety*. Cue-controlled relaxation means you can induce your own relaxation based on repeating certain cue words to yourself. In essence, you are taught to relax and then silently repeat cue words such as "I am relaxed." After enough practice you can relax during mathematics tests. The cue-oriented relaxation techniques have worked with hundreds of students. For a better understanding of test anxiety and how to reduce it, listen to *How to Reduce Test Anxiety*, (available from the Academic Success Press, Inc., P. O. Box 2567, Pompano Beach, Florida 33072; tel: (305) 785-2034 for $12.95 postpaid).

SUMMARY

General test anxiety is a learned behavior developed by having cognitive and/or somatic responses during previous tests. Somatic anxiety occurs in the body; cognitive anxiety occurs in the mind. General test anxiety is a fear of any type of test. Mathematics test anxiety is a subclass of general test anxiety that is specific to one subject area.

Mathematics test anxiety, like general test anxiety, can decrease your ability to perform on tests. Your ability is decreased by blocked memory and a need for leaving the test room before checking all your answers. General and mathematics test anxiety can be overcome by learning cue-controlled relaxation. However, reducing your mathematics test anxiety does not guarantee success on tests. First you have to know the appropriate material to recall during the test and have good test-taking skills. To substantially reduce your test anxiety, follow the instructions presented by the *How To Reduce Test Anxiety* audio cassette tape.

ASSIGNMENT FOR CHAPTER 3

1. Read Reference B — Stress.
2. Listen to *How to Reduce Test Anxiety*.
3. Describe your best short-term relaxation technique from side one of the *How To Reduce Test Anxiety* audio cassette tape. Practice your short-term relaxation technique.

4. Practice the cue-controlled relaxation technique every day until it takes you two minutes or less to relax before taking a test.

NOTES

Chapter 4
Developing Effective Study Management Skills

One of the main problems students have is managing their time. In high school the student's time is managed by teachers and parents. In college, students have more activities (work, social, study) but less time to complete these activities. In fact, when freshman college students are asked their number one reason for poor grades they indicate that they do not have enough time to study. When students are asked how much time they studied per week, most do not have any idea. *Students who do not effectively manage their study time may fail mathematics.*

Developing a Study Schedule

There are two basic reasons for developing a study schedule. *The first reason is to reserve a certain amount of study time per week. This means emphasizing the number of hours a week you are going to study instead of the number of daily study hours.*

How many hours do you study per week? Ten hours, fifteen hours, twenty hours, thirty hours? Without knowing the amount of your study hours per week, you will not know if you are studying at a productive rate. If your goal is to make a "B" average, and with studying fifteen hours a week you make all "B's" on your tests, then the goal has been met. However, if you study fifteen hours a week and make all "D"'s, then you need to increase your study time. By monitoring your grades and weekly hours studied, an adjustment can be made in your study schedule to get the grades you want.

The second reason for developing a study schedule is efficiency. Efficient study means knowing when you are supposed to study and when you are not supposed to study, so you won't be thinking about other things you should be doing when you sit down to study.

The reverse is also true. When doing other things you will not feel guilty about not studying. This is a consequence that most students think other students do not have.

FOR EXAMPLE, you are sitting home on a Saturday afternoon watching a good football game; you start feeling guilty. You have not started studying for that mathematics test on Monday. By having a study schedule you can arrange to study for the mathematics test on Sunday and enjoy the football game on Saturday.

A study schedule is set up for two reasons:

1. To figure the amount of study time you need per week to get the grades you want.

2. To set up peak efficient study times.

Planning Use of Daily Time

To develop a study schedule, locate the "Planning Use of Daily Time" (See **Figure 8**) and review it. Use the "Planning Use of Your Daily Time" as your study schedule.

The best way to develop your study schedule is to begin by filling in all the times you cannot study.

Step 1 — *Fill in all your classes by putting code C (class) in the correct time spaces.* For example, if you have an eight to nine-thirty class, draw a line through the center of the 9:00 box on the study schedule.

Step 2 — *Fill in the time you work (W=work).* Putting in the work times is difficult, since some student's work schedule may change during the week. Some work schedules change every week or every other week. The best way to predict work time is based on the time you worked last week unless you are on a rotating shift. Put down the approximate work times on the study schedule. As your work hours change, revise the study schedule. REMEMBER: The study schedule is mainly looking at the number of hours a week you plan to study. Realize that your work times might change every week but your total weekly work hours usually remain the same.

Step 3 — *Decide the amount of time it takes to eat (E=eat)*; this includes food preparation and cleanup. Keep in mind that the amount of time it takes to eat may fluctuate. So include enough time to prepare the meals, eat the meals, and clean up afterwards. In the study schedule put down approximately when you eat breakfast, lunch, and dinner. Eating also includes time in the student cafeteria. If

"PLANNING USE OF DAILY TIME"

you have an eleven to twelve or one to two lunch break, you might not eat during the entire time. You could be there socializing and eating at the same time. Still put "E" in the study schedule, since the main use of your time is for eating.

Step 4 — *Include your personal hygiene time (PH= personal hygiene).* Some personal hygiene behaviors are taking a bath, washing your hair, or other activities that are required to get ready for school, dates, or work. Personal hygiene varies from minutes to hours a day for college students. Put in the study schedule the usual amount of time spent on personal hygiene. Remember that more time might be spent on personal hygiene during weekends.

Step 5 — *Put down your tutor time (T=tutor).* This is not considered study time. Tutor time is the time you ask the tutor questions about the previous homework assignment. If you have a tutor scheduled or meet weekly with your instructor, put these times in the study schedule.

Step 6 — *Reserve time for family responsibilities on the study schedule (FR=family responsibilities).* Some family responsibilities include taking your child on errands, mowing the lawn, grocery shopping, and taking out the garbage. Also, if you have arranged to take your children some place every Saturday morning, then put it on the study schedule.

Step 7 — *Figure out how much time is spent on cleaning each week (CN = cleaning).* This time can include cleaning your room, house, car, and clothes. Cleaning time usually takes several hours a week. Put in your cleaning time on the study schedule and make sure it is adequate for the entire week.

Step 8 — *Review your sleep patterns for the week (SL=sleep).* Your sleep time will probably be the same from Monday to Friday. On the weekends you might sleep later during the day and stay up later at night. However, if you have been sleeping on Saturdays mornings until ten o'clock for the last two or three years, do not leave time at eight o' clock to study. Be realistic when putting down your sleep time on the study schedule.

Step 9 — *Figure the amount of weekly social time (SC=social time).* Social time includes being with other people, watching TV, or going to church. Social time can be doing nothing or going out and having a good time. Some students mark off the entire weekend for social time. You need to have some social time during the week or you'll burn out. This means you might last one semester. If you study and work too hard without some relaxation, you will not last the school year. Some daily social time is needed but do not over do it. Now put down your social time on the study schedule.

Step 10 — *Recall other time obligations that have not been previously mentioned (O=other)*. Other time obligations may be aspects of your life which you do not want to share with other people. Review the study schedule for any other time obligations. Now write them in.

Now count up all the blank spaces. Each blank space represents one hour. Remember you might have several half blank spaces that represent a half hour. Take that total number, put it on the upper right hand corner of the study schedule and circle it.

Next, figure how many hours to study during the week. The rule of thumb is to study approximately two hours per week for each class hour. If you have twelve real class hours (not counting PD) per week you should be studying twenty to twenty-four hours a week to make A's and B's. Write the amount of time you want to study per week in the upper left hand corner of your study schedule and put a square around it. This is a study contract you are making with yourself.

If the number of contracted study hours is less than the number in the circle, then fill in the times you want to study (S = study). Fill in the best times to study first. If there are any unmarked spaces, use them as back up study time. Now you have the best times to study.

On the other hand, if you want to study fifteen hours a week and only have ten hours of space, then you have to make a decision. Go back over your study schedule codes and locate where you can change some times. If you have a problem locating additional study time, make a priority time list. Then take the hours away from the items with the least priority. Complete the study schedule by putting in your best study times.

FIGURE 8

PLANNING USE OF YOUR DAILY TIME

Master Plan

	7:00	8:00	9:00	10:00	11:00	12:00	1:00	2:00	3:00	4:00	5:00	6:00	7:00	8:00	9:00	10:00	11:00	12:00
MONDAY																		
TUESDAY																		
WEDNESDAY																		
THURSDAY																		
FRIDAY																		
SATURDAY																		
SUNDAY																		

Permission is granted to enlarge FIGURE 8 on a copy machine.

Selecting a Grade as a Course Goal

Determine what grade you want to make in the mathematics course. The grade to write on the study schedule should be an A, B, or C. Do not put down an N, W, X, or F, because these grades mean you will not complete the course. The grade you selected to make in mathematics is now your goal.

Currently you have a study schedule representing the amount of hours of study per week. You also have a course grade goal. After being in the course for several weeks, you will know if you are accomplishing your goal. Should you not be meeting your goal, improve the quality and quantity of your studying or lower your course grade goal.

Selecting a Grade Point Average (GPA) Goal for the Semester

After figuring out what grade you want in the mathematics course, write down the grade point average you want for the semester on your study schedule. Do you want a 4.0 average, (all "A"'s), a 3.0 average, ("B"'s), a 2.5 average,("B"'s and "C"'s), or 2.00 average (all "C"'s). Do not pick any average below 2.00; you will not graduate with a lower average. Be realistic in deciding an overall grade point average.

Developing an Effective Study Plan for the Week

Lastly, develop an effective study plan for the next week and establish weekly study goals. Every Sunday develop a plan to best use your study time for the next week. The first priority is to establish the best time to study math. Math should be studied as soon as possible after each class session. After writing in the Weekly Study Goal Sheet (**Figure 9**) when you will study math for the next week, fill in the remaining study time with your other subjects. For example, write when you will study for your English test and do the history reading assignment. This plan becomes especially important during the week of midterms and final exams.

FIGURE 9

WEEKLY STUDY GOAL SHEET

SUBJECTS	MON.	TUES.	WEDS.	THURS.	FRI.	SAT.	SUN.
Math Course:							

Permission is granted to enlarge Figure 9 on a copy machine.

SUMMARY

You have now completed a study schedule and Weekly Study Goal Sheet, indicating the times to study and the amount of study hours per week. The amount of study hours you contracted with yourself can change based on the grades you want. If you do not receive the type of grades you want in mathematics or overall average during the semester, then increase the quality and quantity of study time. Taking control over your study time can greatly improve your grades.

ASSIGNMENT FOR CHAPTER 4

1. Complete Figure 8 — "Planning Use of Your Daily Time."
2. Select an "A", "B", or "C" as a mathematics course goal.
3. Select between a 2.00 and 4.00 grade point average as an overall semester goal.
4. Complete Figure 9 — "Weekly Study Goal Sheet."

NOTES

Chapter 5
Improving Listening And Note-Taking Skills

Effective Listening

Becoming an effective listener is the foundation for good note-taking. Effective listening can be learned through a set of intricate skills you can learn and practice. To become an effective listener you must prepare yourself physically and mentally.

The physical preparation in becoming an effective listener involves where you sit in the classroom. Sit in the best area to obtain high grades: *The golden triangle of success.* The golden triangle of success begins with seats in the front row and converges to the middle seat in the back row facing the teacher's desk (See **Figure 10**). Students sitting in this area directly face the teacher and have to pay attention to the lecture. Also, there is less tendency for them to be distracted by students outside the classroom or by students making noise within the classroom.

You can hear the instructor better because the instructor's voice is projected to the middle seat in the back row. This means there is less chance of misunderstanding the instructor and you can hear well enough to ask appropriate questions. Thus by sitting in the "golden triangle of success" you can force yourself to pay more attention during class and be less distracted by other students. This is very important for math students because math instructors usually only go over a point once and continue on to the next point. If you miss that point in the lesson then you could be "lost" for the remainder of the class.

The mental preparation of note-taking involves "warming up" before class begins and becoming an active listener. Just like an athlete must "warm up" before a game begins you must "warm up" before taking notes. "Warm up" by:
* ★ reviewing the previous day's notes
* ★ reviewing the reading material
* ★ reviewing the homework, or by

FIGURE 10

THE GOLDEN TRIANGLE OF SUCCESS

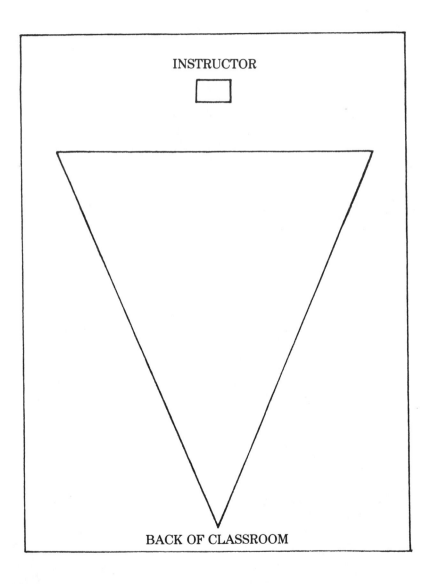

★ preparing questions for the instructor.

This mental "warm up" before the lecture allows you to refresh your memory and prepare pertinent questions, making it easier to learn the new lecture material.

Becoming an active listener is the second part of the mental preparation for note-taking. Some ways to become an active listener include watching the speaker, listening for main ideas, and nodding your head or saying to yourself, "I understand," when agreeing with the instructor.

Don't anticipate what the instructor is going to say or immediately judge the instructor's information before the point is made. This will distract you from learning the information.

Expend energy looking for interesting topics in the lecture. When the speaker discusses information that you need to know, immediately repeat it to yourself to begin the learning process.

REMEMBER. Class time is an intense study period that should not be wasted.

Learning Good Note-Taking Techniques

To become a good note taker requires two basic philosophies. *One philosophy is to be specific in detail. In other words, copy the problems down step by step. The second philosophy is to understand the general principles, general concepts, and general ideas.*

While taking mathematics notes you need to copy each and every step of the problem, even though you may know every step of the problem. While in the classroom you might understand each step, but a week from now you might not remember how to do the problem unless all the steps are written in your notes. In addition, as each step is written down, it is being memorized. *Make sure to copy every step for each problem written on the board.*

There will be times when you get lost while listening to the lecture. You should keep taking notes, though you do not understand the problem. This will provide a reference point for further study. Put a question mark by those steps you do not understand. After class, review the steps you did not understand with the instructor or with a fellow student.

The goal of note-taking is to take the least amount of notes and get the most amount down. This could be the opposite of what most instructors have told you. Some instructors tell you to take everything down. This is not a good note-taking system. It is very difficult to take precise, specific notes and hear what the instructor is saying. Thus, what you need to develop is a note-taking system where you write the least amount possible and get the most amount down while still hearing what the instructor is saying.

Developing Abbreviations

To reduce the amount of written notes, an abbreviation system is needed.

FOR EXAMPLE: When the instructor starts explaining the commutative property you need to write it out the first time. After that, use "COM". You need to develop abbreviations for all the most commonly used words in mathematics.

Figure 11 has a list of abbreviations. Add your own abbreviations to this list. *By using abbreviations as much as possible, you can obtain the same meaning from your notes and have more time to listen to the instructor.*

Utilizing The Modified Two-Column System

Another procedure to save time while taking notes is to stop writing complete sentences. Write your main thoughts in phrases. Phrases are easier to put down and easier to memorize.

FIGURE 11

ABBREVIATIONS AND SYMBOLS

E.G. _____ For Example

CF. _____ Compare, remember in context

N.B. _____ Note well, this is important

.·. _____ Therefore

.·. _ Because

⊃_____ Implies, it follows from this

> _____ Greater than

< _____ Less than

= _____ Equals, is the same

/ _____ Does not equal, is not the same

( ) _____ Parentheses in the margin, around a sentence or group of sentences indicates an important idea.

? _____ Used to indicate you do not understand material

o _____ A circle around a word may indicate that you are not familiar with it; look it up

E_____ Marks important materials likely to be used in exam

1, 2, 3, 4 _____ To indicate a series of facts

D _____ Shows disagreement with statement or passage

REF_____ Reference

ET AL._____ And others

BK_____ Book

P_____ Page

ETC. _____ And so forth

V._____ See

V.S._____ See above

VS_____ Against

SC_____ Namely

SQ_____ The following

Comm._____ Commutative

Dis._____ Distributive

A.P.A. _____ Associative Property of Addition

A.I. _____ Additive Inverse

I.P.M. _____ Identity Property of Multiplication

One of the best mathematics note-taking systems is the modified two column system (See **Figure 12**). This note-taking system divides the note page into three areas. The first area is the margin left of the red line and is used for putting down key ideas and key words. Key words or key ideas represent the main focal points of the lecture such as absolute value, important formulas, and the associative property rule. If possible, *write down the key words and key ideas during the lecture.*

FIGURE 12

MODIFIED TWO-COLUMN SYSTEM

KEY IDEAS WORDS	DISCUSSION OF RULES/ NOTES	EXAMPLES
		$5(W+3) + 3(W+1) = 14$
	Distributive Property	$5W + 15 + 3W + 3 = 14$
	Commutative Property	$5W + 3W + 15 + 3 = 14$
Associative Property	Associative Property	$(5W + 3W) + (15 + 3) = 14$
	Addition of like terms	$8W + 18 = 14$
Add	Addition property of	
opposite of	equality	$8W + 18 - 18 = 14 - 18$
18	Addition + additive	
	inverse	$8W = -4$
	Multiplication property	
	of equality	$1/8 (8W) = 1/8 (-4)$
	Reduce	$W = -4/8$
	Answer	$W = -1/2$
	SOLVING EQUATIONS:	CHECKING ANSWERS:
	Put variable on one side	See if seems right
	Put numbers on other side	Put answer back into
	Simplify	equation
		solve equation

The next note-taking section is Discussion of Rules. In this section write down and explain the important algebraic rules that correspond to the problem. Also, add any other important notes by writing short phrases and abbreviations. Put question marks by the material which is not understood.

The example section (far right) in the note system is for writing out the problems. Each problem put on the board or discussed in class should be written down step by step. If you get confused by a step, then put a question mark by it for additional study. If possible, reference the examples to a page number in the text book.

Knowing When and When Not to Take Notes

To become a better note taker you need to know when to take notes and when not to take notes. The instructor will give cues that indicate what material is important. Some cues given are:
* ★ deviant facts or ideas
* ★ writing on the board
* ★ indicating that certain problems will be on the test
* ★ summarizing
* ★ pausing
* ★ repeating statements
* ★ enumerating such as "1, 2, 3" or "A, B, C"
* ★ when an instructor works several examples of the same type of problem on the blackboard
* ★ when the instructor says, "This is a tricky problem. Most students will miss it." For example, 5/0 is undefined instead of zero.
* ★ when the instructor says, "This is the most difficult step in the problem"
* ★ when the instructor indicates that certain types of problems will be on the test, such as coin or age word problems

You need to learn the cues your instructor gives indicating important material. If you are in doubt about the importance of the class material, do not hesitate to ask the instructor about its importance.

While taking notes, math material can become confusing. At this point take as many notes as possible. Don't give up on note-taking. As you take notes on confusing problem steps, skip lines; then go back and fill in information that clarifies your misunderstanding of the steps in question. Ask the tutor or instructor for help with the uncompleted problem steps and write down the reasons for each step in the space provided.

"USING A TAPE RECORDER"

Using a Tape Recorder

If you have a math instructor who lectures too fast or if you cannot get all the information down while listening to the lecture, use a tape recorder. To ensure success, the tape recorder must have a tape counter. The tape counter displays a number indicating the amount of tape to which you have listened. When you find you're in an area of confusing information, write the beginning and ending tape counter number in the left margin of your notes. When reviewing your notes, the tape number will be a reference point for obtaining information to work the problem. You can also reduce the time it takes to listen to the tape by using the pause button to stop the recording of unnecessary material.

Asking Questions

To obtain the most from a lecture, you must ask questions in class. By asking questions, you improve your understanding of the material and decrease your homework time. By not asking questions, you create for yourself unnecessary confusion during the remainder of the class period. Also, it is much easier to ask questions in class about potential problems in the homework rather than spending hours trying to figure out the problems on your own at a latter time.

If you are shy about asking questions in class, then write down the questions and read them to your instructor. If the instructor seems confused about the questions, tell him you will talk to him after class. To encourage yourself to ask questions, remember:

(1) You have paid for the instructor's help;

(2) Five other students probably have the same question, and

(3) The instructor needs feedback on his teaching to help the class learn the material.

Recording Each Problem Step

My last suggestion in note-taking is to *record each step of every problem written or verbally explained.* By recording each problem step, you begin overlearning how to work the problems; this will increase your problem-solving speed during future tests. Also, if you get stuck on the homework, you will have complete examples to review. The major reason for this procedure is to understand how to do the problems while the instructor is explaining them — instead of trying to remember unwritten steps. Writing down each step of a problem discussed by the instructor may seem time consuming. But it pays off during homework and test time.

Listening and Learning

Some students think listening to the instructor and taking notes is a waste of valuable time. Students too often sit in class and use a fraction of their learning ability. Class time should be considered a valuable study period where you can listen, take notes, and learn at the same time. One way to do this is by memorizing important facts when the instructor is talking about material you already know. Another technique is to repeat back to yourself the important concepts right after the instructor says them in class. *Using class time to learn mathematics is an efficient learning system.*

Reworking Your Notes

The note-taking system does not stop when you leave the classroom. *As soon as possible after class, rework your notes.* You can rework the notes between classes or as soon as you get home. This is especially important for learning disabled students.

REMEMBER: Most forgetting occurs right after learning the material. You need to rework the notes to obtain the most value. Do not wait two weeks to rework the notes. You probably will not understand what was written.

The following are important steps in reworking your notes:

Step 1 — *Rewrite the material you cannot read or will not be able to understand in two weeks.* If you do not rework your notes it will be frustrating to come across important material that is illegible when you start studying for a test. Another benefit of rewriting the notes is the immediate learning of the new material instead of taking more time to learn the material a few days later.

Step 2 — *Fill in the gaps.* Most of the time when you are listening to the lecture, you can't write everything down. It is almost impossible to write everything down even if you know shorthand. Locate the portions of your notes which are incomplete. Fill in the concepts which were left out. In the future, skip two or three lines on your notebook page for known lecture gaps.

Step 3 — *Add additional key words and ideas in the left hand column.* These key words or ideas were the ones not recorded during the lecture. FOR EXAMPLE: You did not know you should add the opposite of eighteen to solve a problem. Put additional important key words and ideas in the notes; these are the words that will improve your understanding of mathematics.

Step 4 — *Reflection and synthesis.* Once you have finished going over your notes, review the major points in your mind. Synthesize your notes by combining your previous knowledge with what you have learned today.

SUMMARY

Effective listening is the first step to excellent note-taking. The effective listener knows where to sit in the classroom and understands good listening techniques.

The goal of note-taking is to write the least amount possible to record the most information. This allows you to enhance your ability to listen to the lecture and increase your learning potential in the classroom.

Re-work your notes as soon as possible after class. If you wait too long to review your notes, then you might not understand them and it will be more difficult to learn them. Re-working your notes will improve your understanding of mathematics — and your grades.

Learning disabled students might need special help with listening skills and note-taking. Refer to Reference I for specific note-taking suggestions.

ASSIGNMENT FOR CHAPTER 5

1. Review Figure 11 — Abbreviations.
2. Review Figure 12 — Modified Two-Column System.
3. How can you become an effective listener?

4. Why do you need to copy down each step of the math homework?

5. What abbreviations do you use in your math notes?

6. What cues do math instructors give to indicate important test information?

7. How does asking questions in math class decrease the time you will spend on homework?

8. Do you have a classmate with whom to compare math notes?

9. How are you going to improve your math note-taking system?

Chapter 6

Improving Reading and Homework Techniques

Reading a Math Textbook

Reading a mathematics textbook is more difficult than reading other textbooks. The best process for reading a mathematics textbook is different from the typical way students are taught reading in high school. In high school, students are taught to read quickly or skim the material. If you did not understand a word, you were to keep on reading. The reason for continuing to read was to pick up the unknown words and their meaning from context.

This reading technique may work with your other classes, but using it in your mathematics course will be totally confusing. By skipping some major concept words or bold printed words, you will not understand the mathematics textbook or be able to do the homework. Reading a mathematics textbook takes more time and concentration than your other textbooks.

Understanding Reading Material

There are several appropriate steps in reading a mathematics textbook:

Step 1: — *Skim the assigned reading material.* Skim the material to get the general idea about the major topics. You do not want to learn the material at this time; you want to get an overview of the assignment. FOR EXAMPLE: Skimming will allow you to see if problems presented in one chapter section are further explained in subsequent chapter sections.

Step 2: — *As you skim the chapter, circle the new words that you do not understand.* If you do not understand these new words after reading the assignment, then ask the instructor for help. Skimming the reading assignments should only take five to ten minutes.

Step 3: — *Put all your concentration into reading.* REMEMBER: Reading a mathematics textbook is very difficult. It might take you half an hour to read and understand one page. Do not skip any of the reading assignment.

Step 4: — *When you get to the examples go through each step.* If the example skips any steps, make sure you write down each one of those steps in the textbook for better understanding. Later on, when you go back and review, the steps are already filled in. You will understand how each step was completed. Also, by filling in the extra steps, you are starting to over-learn the material for better recall on future tests.

Step 5: — *Mark the concepts and words that you do not know.* Maybe you marked them the first time while skimming. If you understand them now, erase the marks. If you do not understand the words or concepts, then re-read the page or look them up in the glossary. Try not to read any further until you understand all the words and concepts.

Step 6: — *If you do not clearly understand some words or concepts, develop your own glossary in the back of your notebook and ask the instructor for a better explanation.* You should know all the words and concepts in your glossary before taking the test.

Step 7: — *If you do not understand the material, follow these eight steps, one after the other, until you do understand the material:*

A: Go back to the previous page and re-read the information to maintain a train of thought.

B: Read ahead to the next page to discover if any additional information better explains the misunderstood material.

C: Locate and review any diagrams, examples or rules that explain the misunderstood material.

D: Read the misunderstood paragraph(s) several times aloud to better understand their meaning.

E: Refer to your math notes for a better explanation of the misunderstood material.

F: Refer to another math text book or use math computer software that expands the explanation of the misunderstood material.

G: Define exactly what you do not understand and call your study buddy for help.

H: Contact your math tutor or math instructor for help in understanding the material.

Hopefully you will not have to use all eight steps when coming across material you do not understand.

Step 8: While reading the textbook, *highlight the material that is important to you*. However, do not highlight over 50% of a page because the material is not being narrowed down enough for future study. Especially highlight the material that is also discussed in the lecture. *Material discussed both in the textbook and lecture usually appears on the test*. The purpose for highlighting is to emphasize the important material for future study.

By using these reading techniques, you have narrowed down the important material for learning. You have skimmed the textbook to get an overview of the assignment. You have carefully read the material and highlighted the important parts. You then developed a glossary of unknown words or concepts which further narrowed down the material and subsequently became the most important study material.

In summary, the highlighted material should be reviewed before doing the homework problems and the glossary has to be learned 100 percent before taking the test.

Establishing Study Period Goals

Before beginning your homework, establish goals for the study period. Don't just do the homework problems.

Ask yourself this question: "What am I going to do tonight to become more successful in math?"

By setting up short-term homework goals and accomplishing them, you feel more confident about mathematics. This also improves your self-esteem and helps move you toward becoming a more internally motivated student. You have set tasks which you have completed.

Study period goals are set up either on a *time-line basis or an item-line basis*. A time-line basis is studying math for a certain amount of time. FOR EXAMPLE: You may want to study math for an hour, then switch to another subject.

An item-line basis is studying math until you have completed a certain number of homework problems. For instance, you might set a goal to study math until you have completed all the odd problems in the chapter review.

No matter what homework system you use, remember this important rule: *Always finish a homework session by understanding a concept or doing a homework problem correctly.* Do not end a homework session with a problem you cannot complete. You will lose confidence, since all you'll think about is the last problem you couldn't solve instead of the 32 problems you did solve correctly. If you did quit on a problem you couldn't solve, return and re-work problems you have done correctly. *Don't end your study period with a problem you couldn't complete.*

Do Your Homework : Ten Steps

Doing your homework can be frustrating or rewarding. Most students try to do their homework without any type of preparation, become frustrated and stop studying. To improve your homework success, follow these ten steps:

Step 1: — Review the textbook material that relates to the homework. A proper review will increase the chances of successfully completing your homework. If you get stuck on a problem, you will have a better chance of remembering the location of similar problems. If you do not review prior to doing your homework, you could get stuck and not know where to find the help in your textbook. To be successful in learning the material and in completing homework assignments, first review your textbook.

Step 2: — Review your lecture notes that relate to the homework. If you could not understand the explanation of the textbook on how to complete the homework assignment, then review your notes. Reviewing your notes will give you a better idea on how to complete your homework assignment.

Step 3. — Do your homework as neatly as possible. Doing your homework neatly has several benefits. When approaching your instructor about problems with your homework, he or she will be able to understand your previous attempts to solve the problem. The instructor will easily locate the mistakes and show you how to correct the steps without having to decipher your handwriting. Another benefit is that when you review for midterm or final exams, you can quickly re-learn the homework material without having to decipher your own writing. Neatly prepared homework can help you now and in the future.

Step 4. — When doing your homework, write down every step of the problem. Even if you can do the step in your head, write it down anyway. This will increase the amount of homework time, but you are over-learning how to solve problems which improves your memory. By doing every step, it is an easy way to memorize and understand the material. Another advantage is that when you re-work problems that you did wrong it is easy to review each step to find the mistake. In the long run, *doing every step of the homework will save you time and frustration.*

Step 5. — Understand the reasons for doing each set of the homework problems. Don't get into the bad habit of memorizing how to do problems without knowing the reasons for each step. Many students are smart enough to memorize procedures required to complete a set of homework problems. However, when similar homework problems are presented on a test, the student cannot solve the problems. To avoid this dilemma, keep reminding yourself about the rules, laws, or properties used to solve problems. Here's an example.

PROBLEM: $2(a + 5) = 0$.

What property allows you to change the equation to $2a + 10 = 0$? ANSWER: The distributive property.

Once you know the correct reason for going from one step to another in solving a math problem, then you can answer any problem requiring that property. *Students who memorize how to do problems — instead of understanding the reasons for correctly working the steps — will eventually fail their math courses.*

Step 6. — If you do not understand how to do a problem follow these steps:

A. Review the textbook material that relates to the problem.

B. Review the lecture notes that relate to the problem.

C. Review any similar problems, diagrams, examples or rules that explain the misunderstood material.

D. Refer to another math textbook, math computer program software, or math video to obtain a better understanding of the material.

E. Call your study buddy.

F. Skip the problem and contact your tutor or math instructor as soon as possible for help.

Step 7. — Always finish your homework by successfully completing problems. Even if you get stuck, go back and

"SOLVING WORD PROBLEMS"

successfully complete previous problems before quitting. You want to end your homework assignment positively with feelings of success.

Step 8. — After finishing up your homework assignment, recall to yourself or write down the most important learned concepts. Recalling this information will increase your ability to learn these new concepts. Additional information about Step 8 will be discussed later in this chapter.

Step. 9. — Make up note cards containing hard-to-remember problems or concepts. Note cards are an excellent way to review material for a test. More information on the use of note cards as learning tools is discussed later in this chapter.

Step 10. — Getting behind in mathematics homework is academic suicide. As mentioned in Chapter One, mathematics is a linear learning process. If you get behind in that learning sequence, it will be very difficult to catch up because each topic builds on the next. It will be like going to Spanish class without learning the last set of vocabulary words. The teacher will be talking to you, using the new vocabulary, but you won't understand what is being said.

To keep up with your homework, it is necessary to complete the homework every school day and sometimes even on weekends. Doing your homework one-half hour a day for two days in a row is better than one hour every other day. If you have to get behind in one of your courses, make sure it is not math. Fall behind in a course that does not have a linear learning process; i.e., psychology.

REMEMBER: Getting behind in math homework is the fastest way to fail the course.

Solving Word Problems

The most difficult homework assignment for most math students is working story/word problems. Solving word problems requires excellent reading comprehension and translating skills. Students have difficulty substituting English terms for Algebraic symbols and equations. Once an equation is written, it is usually easily solved. To help you solve word problems follow these ten steps:

1. Read the problem three times. Read the problem fast the first time as a scanning procedure. As you are reading the problem the second time, answer these three questions:

    a. What is the problem asking me (usually at the end of the problem)?

    b. What is the problem telling me that is useful (cross out unneeded information)?

    c. What is the problem implying (usually something you have been told to remember)?

    Read the problem a third time to check that you fully understand its meaning.

2. Draw a simple picture of the problem to make it more real to you, (e.g., a circle with an arrow can represent travel in any form — by train, by boat, by plane, by car, or by foot).

3. Make a table of information and leave a blank space for the information you are not told.

4. Use as few unknowns in your table as possible. If you can represent all the unknown information in terms of a single letter — do so! When using more than one unknown use a letter that reminds you of that unknown. Then write down what your unknowns represent. This eliminates the problem of assigning the right answer to the wrong unknown. Remember you have to create as many separate equations as you have unknowns.

5. Translate the English terms into an Algebraic equation using the list of terms in **Figures 13** and **14**. Remember the English terms are sometimes stated in a different order than the algebraic terms.

6. Immediately retranslate the equation as you now have it written back into English. The translation will not sound like a normal English phrase but the meaning should be the same as the original problem. If the meaning is not the same, then the equation is incorrect and needs to be rewritten. Rewrite the equation until it means the same as the English phrase.

7. Review the equation to see if it is similar to equations from your homework and if it makes sense.

8. Solve the equation using the rules of Algebra. Remember whatever is done to one side of the equation must be done to the other side of the equation. The unknown must end

FIGURE 13

TRANSLATING ENGLISH TERMS
INTO ALGEBRAIC SYMBOLS

Sum	+
Add	+
In addition	+
More than	+
Increased	+
In Excess	+
Greater	+
Decreased by	−
Less than	−
Subtract	−
Difference	−
Diminished	−
Reduce	−
Remainder	−
Times as much	×
Percent of	×
Product	×
Interest on	×
Per	/
Divide	/
Quotient	/
Quantity	( )
Is	=
Was	=
Equal	=
Will be	=
Results	=
Greater than	>
Greater than or equal to	≥
Less than	<
Less than or equal to	≤

up on one side of the equation by itself. If you have more than one unknown then use the substitution or elimination method to solve the equations.

FIGURE 14

TRANSLATING ENGLISH WORDS
INTO ALGEBRAIC EXPRESSIONS

English Words	Algebraic Expressions
Ten more than x	$x + 10$
A number added to 5	$5 + x$
A number increased by 13	$x + 13$
5 less than 10	$10 - 5$
A number decreased by 7	$x - 7$
Difference between x and 3	$x - 3$
Difference between 3 and x	$3 - x$
Twice a number	$2x$
Ten percent of x	$.10x$
Ten times x	$10x$
Quotient of x and 3	$x/3$
Quotient of 3 and x	$3/x$
Five is three more than a number	$5 = x + 3$
The product of two times a number is 10	$2x = 10$
One half a number is 10	$x/2 = 10$
Five times the sum of x and 2	$5(x + 2)$
Seven is greater than x	$7 > x$
Five times the difference of a number and 4	$5(x - 4)$
Ten subtracted from 10 times a number is that number plus 5	$10x - 10 = x + 5$
The sum of 5x and 10 is equal to the product of x and 15	$5x + 10 = 15x$
The sum of two consecutive integers	$(x) + (x + 1)$
The sum of two consecutive even integers	$(x) + (x + 2)$
The sum of two consecutive odd integers	$(x) + (x + 2)$

9. Look at your answer to see if it makes common sense. FOR EXAMPLE: If tax was added to an item, it should cost more or if a discount was applied to an item, it should cost less. Is there more than one answer? Does your answer match the original question? Does your answer have the correct units?

10. Put your answer back into the original equation to see if it is correct. Note: if one side of the equation equals the other side of the equation, then you have the correct answer. If you don't have the correct answer then go back to step five.

The most difficult part of solving word problems is translating part of a sentence into Algebraic symbols and then into Algebraic expressions. **Figures 13** and **14** list Algebraic symbols and expressions which will help you solve word problems.

Factoring

Another common area of difficulty for algebra students is factoring. Math instructors have indicated that not knowing how to factor is one of the most common causes of failure in

FIGURE 15

MULTIPLYING SPECIAL PRODUCTS

I.	$-(x - y)$	$=$	$-x + y$ often written and $y - x$
II.	$a(b + c)$	$=$	$ab + ac$
III.	$(a + b)\,(m + n)$	$=$	$(a + b)m + (a + b)n$
IV.	$(a + b)^2$	$=$	$a^2 + 2ab + b^2$
V.	$(a - b)^2$	$=$	$a^2 - 2ab + b^2$
VI.	$(a + b)\,(a - b)$	$=$	$a^2 - b^2$
VII.	$(x + g)\,(x + h)$	$=$	$x^2 + (g + h)x + gh$
VIII.	$(gx + r)\,(sx + t)$	$=$	$gsx^2 + (gt + rs)x + rt$
IX.	$(a + b + c)^2$	$=$	$a^2 + b^2 + c^2 + 2ab + 2ac + 2bc$
X.	$(a + b)\,(a^2 - ab + b^2)$	$=$	$a^3 + b^3$
XI.	$(a - b)\,(a^2 + ab + b^2)$	$=$	$a^3 - b^3$
XII.	$(a + b)^3$	$=$	$a^3 + 3a^2b + 3ab^2 + b^3$
XIII.	$(a - b)^3$	$=$	$a^3 - 3a^2b + 3ab^2 - b^3$
XIV.	$(a - b + m)\,(a - b - m)$	$=$	$a^2 - 2ab + b^2 - m^2$

algebra. Read very carefully the following suggestions on improving your ability to factor.

Know these important special products (results of multiplying two or more variables) in **Figure 15** to assure success in factoring, which is an important part of any algebra course. These common products occur so often that they need to be memorized so you can write the results without doing the multiplication. These special products are as important to basic algebra as the multiplication tables are to arithmetic. You must know these patterns both forward and backwards, so that when given the factors you can state the product and when given the product you can state the factors. The statements in **Figure 15** can be verified by doing the multiplication. If you are in a beginning or intermediate algebra class complete the multiplication of each statement in **Figure 15** to build your speed and confidence.

Using Note Cards to Improve Your Scores

After completing your homework problems, a good learning technique is to make note cards. Note cards are 3" x 5" index cards which contain information that is difficult to learn or material you think will be on the test. On the front of the note card write a math problem or information that you need to know. On the back of the note card write how to work the problem or an explanation of pertinent information. FOR EXAMPLE: If you are having difficulty remembering the rules for multiplying positive and negative numbers, then you would write some examples on the front of the note card with the answers on the back.

In this fashion you can glance at the front of the card, repeat to yourself the answer, and check yourself with the back of the card. Make note cards on important information you might forget. Every time you have five spare minutes, pull out your note cards and review them.

Working With a Study Buddy

You need to have a study buddy when doing your homework. A study buddy is a friend who is taking the same course. You can call your study buddy when you get stuck on your homework. Do not sit for half an hour or hour trying to work one problem; it will destroy your confidence and waste valuable time.

Think how much you could have learned by trying the problem for fifteen minutes and then calling your study buddy for help. Spend, at the maximum, fifteen minutes on one problem before going on to the next problem or calling your study buddy. A study buddy can improve your learning while completing the homework.

Recalling What You Have Learned

After finishing your homework, close the textbook and try to remember what you have learned. Ask yourself this question, "What major concepts did I learn tonight?"

Recall for about three to four minutes the major points of the assignment, especially the areas you had difficulty understanding. Since most forgetting occurs right after learning the material, this short review will help you retain the new material.

The final study step is reading ahead. If you read ahead, do not expect to understand everything. Read ahead two or three sections and put question marks by the material you do not understand. When the instructor starts discussing that material have your questions prepared and take good notes. Or if it is getting late in the lecture period and the instructor has not covered a concept which will be on the homework, raise your hand and ask questions about it. Reading ahead will take more time and effort, but it will better prepare you for the lectures.

SUMMARY

You will better understand mathematics if you:
★ Do your homework neatly
★ Write down every step of a problem
★ Get a study buddy
★ Review what you learned from the homework, and
★ Read ahead in the textbook.
★ If you have a learning disability, then read Reference I for more information on reading a math textbook.

ASSIGNMENT FOR CHAPTER 6

1. Review **Figure 13** — Translating English Words into Algebraic Symbols and **Figure 14** — Translating English Words into Algebraic Expressions.

2. Review **Figure 15** — Factoring Special Products.

3. How is reading a math textbook different from reading other textbooks?

4. What is the best procedure to use when reading a math textbook?

5. What do you need to do before starting the math homework?

6. Why do you need to write down every problem step while doing the homework?

7. What happens on math tests when you only memorize how to do the math homework?

8. How can note cards be used to improve math test scores?

9. What are the steps to use to solve word problems?

10. What should you do after finishing your math homework?

NOTES

Chapter 7

Creating A Positive Study Environment

Choosing a Place to Study

Creating a positive study environment can improve the quality of studying. Choosing an appropriate study place may seem trivial, but it can significantly enhance learning. When you start studying, pick a spot at home and at school. While studying in your home, pick one place, one chair, one desk or table as your study area. If you use the kitchen table, pick one chair — preferably one which you do not use during dinner. Now call this chair, "my study chair." If you study in the student cafeteria, use the same table each time. Do not use the table at which you play cards or eat.

By studying at the same place each time, a conditioned response will be formed. From then on, when you sit down at your study place your mind will automatically start thinking about studying. This conditioned response decreases your "warm up" time. "Warm up" time is how long it takes to actually begin studying after you sit down.

Another aspect of the study environment pertains to the degree of quiet you need for studying. In most cases, a totally quiet room is not necessary. But if you can only study with total silence, be careful when selecting your study place and time. Most students can study with a little noise, especially if it is a constant sound, like a mellow radio station. In fact, some students keep on a mellow radio station or a fan to drown out other noises. However, do not turn on the TV to drown out other noises or to listen to while studying. That will not work! For most efficient studying, select a study area where you can control the noise level.

Your study environment should be surrounded by signs that tell you to study. One sign should be your study schedule. Attach your study schedule to the inside flap of your notebook and place

"CHOOSING A TIME AND PLACE TO STUDY"

another copy in your study place at home. Place your study goals and the rewards for achieving those goals where they can be seen easily. Do not post pictures of your girl friend, boy friend, bowling trophies, fishing trophies, or other items in your study area which could be distracting. Post pictures indicating your goals after graduation. If you want to be a nurse, doctor, lawyer, or businessman, post pictures that represent these goals. The study area should reinforce your educational goals.

When sitting down to study, the tools of your trade are required: pencils, paper, notebook, textbook, calculator. Anything you might need should be within grabbing distance. In this way, when you need something you can reach for it, instead of getting up all the time. The problem with getting up is not just the time it takes to get the item, but the time it takes to "warm up" again. After getting milk and cookies and sitting down, it takes another four to five minutes to continue studying.

Studying Subjects in Order of Difficulty

When studying, arrange your subjects in the order of difficulty. In other words, *start with your most difficult subject — which is usually math — and work toward your easiest course.* By studying your most difficult subject first, you are more alert and better motivated to complete the work before continuing on to easier courses which may also be more interesting to you. If you study math last you will probably tire easily, become frustrated, and may quit. However, you are less likely to quit when you study a subject that interests you. REMEMBER: Study math first!

Mixing the Order of Study

Another approach to improving the quality of your study is to mix up the order of studying different subjects. For example, if you have English, accounting, and math to study, then study them in the following order: (1) math, (2) English, and (3) accounting. By studying the subjects in this order, one part of your brain can rest, after studying math, while the other part of your brain is studying English. Now your mind is "fresh" when you study accounting.

Deciding When to Study

Deciding when to study different types of material is part of developing a positive study environment. Your study material can be divided into two separate types. One type is new material and the other type is material that has already been learned. The best time to review the material you have already learned is right before going to sleep. By reviewing it the night before, you will have less brain activity and less physical detractors that would prevent you from recalling the material the next day. FOR EXAMPLE: If you have an eight o'clock test the next day, you should review the material the night before. If you have a ten o'clock test the next day, review the material the night before and the day of the test. Reviewing is defined as reading the material to yourself. You also might review a few problems you've already solved to keep your mind ready, but *do not try to learn any new material the night before the test.*

Learning New Material

Learning new material should be conducted during the first part of the study period. Do not learn new material the night before a test. You'll be setting yourself up for test anxiety. If you try to cram the procedures to solve different types of equations or new ways to factor trinomials you will end up in a state of confusion, especially if you have major problems in learning the new equations or factoring. The next day you will remember not being able to solve the equation or factor the trinomials; this could distract you on the test.

Most students get tired after studying for several hours or before going to bed. If you are tired and try to study new material, it becomes more difficult to retain. It takes more effort to learn new material when you are tired than it does to review old material. When you start getting tired of studying, the best tactic is to begin reviewing previously learned material.

Finding the Most Efficient Time to Study Math

The most efficient time to study is as soon as possible after the mathematics class. Psychologists indicate that most forgetting occurs right after learning the material. In other words, you are going to forget most of what you have learned in the first hour after class. To prevent this mass exodus of knowledge you need to recall some of the lecture material. The easiest way to recall the lecture is to re-work your notes. Reviewing your notes will increase your ability to recall the information and make it easier to understand the homework assignments.

Choosing Between Mass and Distributive Learning

There are two different types of learning processes: "mass

learning" and "distributive learning." Mass learning is learning everything at one time. Distributive learning is studying the same amount of time as mass learning but with study breaks.

EXAMPLE:

Mass Learning — you would study three hours in a row without taking a break, then quit studying for the night.

Distributive Learning — you would study for about fifty minutes with a ten minute break, study for fifty more minutes with a ten minute break, and finish with sixty minutes of studying before stopping.

Psychologists have discovered that learning decreases if you do not take study breaks. *Use the distributive learning procedures to study mathematics.*

If you have studied for only fifteen or twenty minutes and feel you are not retaining the information or your mind is wandering, take a break. If you continue to force yourself to study, you will not learn the material. After taking a break, return to studying.

If you still cannot study after taking a break, review your purpose for studying and your educational goals. Think about what is required to graduate. It will probably come down to the fact that you will have to pass math. Think about how studying mathematics today will help you pass the next test; in turn, this will increase your chances of passing the course and graduating. At this point, write down three positive statements about yourself and three positive statements about studying on an index card. Look at this index card every time you have a study problem.

SUMMARY

A positive study environment can improve your mathematics grades. Establishing several appropriate study places can increase your learning potential. Using distributive learning and studying new and old material at the appropriate times can improve your learning skills. For additional information read Reference C (Ten Steps to Improving Your Study Skills) and Reference D (Suggestions to Teachers for Improving Student Math Study Skills).

ASSIGNMENT FOR CHAPTER 7

1. Read Reference C — Ten Steps to Improving Your Study Skills.
2. Read Reference D — Suggestions to Teachers For Improving Student Math Study Skills.
3. How can you improve your study environment?

4. In what order do you need to study your courses?

5. When should you learn new material?

6. When is the most efficient time to study math?

7. What should you do when you can't study?

Chapter 8

Learning Critical
Memory Techniques

Learning is the process of achieving competency. The three ways of learning are:

* ★ conditioning
* ★ thinking
* ★ and a combination of conditioning and thinking

Learning by Conditioning and Thinking

Conditioning is learning things with a maximum of physical and emotional reaction and a minimum of thinking.

FOR EXAMPLE: Repeating facts to yourself or practicing on a typewriter. Conditioned learning uses rote memory and does not involve thinking about what it means.

Thinking is defined as learning with a maximum of thought and a minimum of emotional and physical reaction. Learning by thinking means you learn by observing, processing, and understanding the material. The most successful way to learn is to combine thinking and conditioning. The best learning combination is to learn by thinking first and conditioning second.

Processing and Storing Information

Memory is different from learning; it requires reception, storage, and retrieval of information. The reception of the information is through the sensory register. Sensory input into the sensory register is usually through touching, seeing and hearing. The sensory register is the first stage of memory that briefly holds an exact image of each sensory experience until it can be processed. If the information is not processed immediately, it is forgotten. The sensory register helps us go from one situation to the next without cluttering up our minds with trivial information.

Short-Term Memory

Information is stored in the brain in short-term memory or long-term memory. Examples of short-term memory are:

★ Looking up a telephone number in the directory and remembering it long enough to dial, then forgetting it immediately

★ Learning the name of a person at a large party or in a class but forgetting it completely within a few hours

★ Cramming for a test and forgetting most of it before taking the test

Remembering something for a short time is not hard to do. By conscious effort you can remember mathematics laws, facts, and formulas from the sensory register and put them in short-term memory. You can recognize them and register them in your mind as something to remember for a short time. When you are studying mathematics you can tell yourself the distributive property is illustrated by a(b+c) = ab + ac. By deliberately telling yourself to remember that fact, you can remember it because you have put it in short-term memory to be taken out when needed. Short-term memory is stored in the front of the mind. These memories can be recalled easily when needed.

The amount of information you can keep in short-term memory is comparatively small. You may be able to tell yourself to remember one phone number or a few formulas but not ten formulas. To remember more facts or ideas, especially at test time, a better system than short-term memory is required. Psychologists have found that short-term memory cannot hold an unlimited amount of information. Futhermore, items placed into short-term memory usually fade fast, like the telephone number you learned just long enough to dial. Short-term memory is useful in helping you concentrate on a few things at a time but is not the best way to learn mathematics.

Long-Term Memory

Long-term memory is a store-house of material that is retained for long periods of time. Long-term memory is not trying harder and harder to remember more and more unrelated facts or ideas. Basically, long-term learning is organizing your short-term memories into meaningful information, thinking about them, comprehending their meaning, and mentally rehearsing them. Without rehearsing the information, it will not be processed into long-term memory.

The main problem students face is converting learned material from short-term memory into long-term memory. Putting mathematics information into long-term memory is not accomplished by just doing the homework. You need to develop effective concentration techniques to place important material into long-term memory.

The last step in the memory process is retrieving the information. This retrieval process becomes critical during tests. Retrieval can be blocked by insufficient processing into long-term memory by test anxiety or by poor test-taking skills. Ways to decrease test anxiety were discussed in Chapter Three and appropriate test-taking procedures will be explained in Chapter Nine.

Understanding the stages of memory will help you answer this common complaint about mathematics: "How can I understand the procedures to solve a math problem one day and forget how to solve a similar problem two days later?"

There are two good answers to this complaint. After learning how to solve the problem, the process was not rehearsed enough times to enter your long-term memory. A second answer is that you did get the information into long-term memory, but the information was not reviewed enough and was forgotten. Either answer represents the results of students not rehearsing or reviewing their recently learned math material. See **Figure 16** for an understanding of the Stages of Memory.

Converting Material From Short-term to Long-term Memory

There are different types of concentration and learning techniques to help convert information into long-term memory.

Having a positive attitude about studying will help you concentrate and improve retention. This does not mean you have to like studying mathematics but you at least need a positive attitude about learning the material. Look at studying as an opportunity to learn rather than an unpleasant task. Tell yourself that you can learn the material and that it will help you pass

"CONVERTING MATERIAL FROM SHORT-TERM TO LONG-TERM MEMORY"

the course and graduate.

Use the "full mind concept" to decrease distractions. Imagine that your mind is completely filled with thoughts of learning mathematics and other distracting thoughts can't enter. Your mind only has one way input and output which only responds to thinking about mathematics when you are doing homework or studying.

Improve your study concentration by counting the number of distractions for each study session. Place a sheet of paper by your book; when you catch yourself not concentrating put the letter "C" on the sheet of paper. This will remind you to concentrate and get back to work. Count up the number of "C's" after each study period and watch the number decrease.

Studying with a pen or highlighter can improve your concentration. Placing the pen or highlighter in your hand and using it will force you to concentrate more.

After reading or studying your notes or textbook for a while, *write down in your own words the main thought.* The secret to this technique is the activity and expectation of learning.

Being selective in your learning will improve your memory. Decide on the facts you need to know and the ones you can ignore.

FIGURE 16

STAGES OF MEMORY

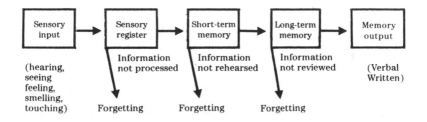

Narrow down information into laws and principles that can be generalized. Learn the laws and principles 100 percent.

Organizing the material will help you learn. Learn and memorize ideas and facts in clusters or groups. Put the material to be learned into categories. Do not learn isolated facts.

Recitation is one of the best ways to get important information into long-term memory. Say facts and ideas out loud; this improves your ability to think and remember. If you cannot recite out loud, recite the material to yourself, emphasizing the key words. Writing and reciting the material at the same time is an excellent way to learn. After you correctly recite the material back to yourself one time, do it five or ten more times to overlearn it.

Once you have worked out a homework problem correctly, do similar problems five times more for over-learning to occur.

Use mental pictures or diagrams to help you learn the material. Mental pictures and actual diagrams involve 100 percent of your brain power. Picture the steps to solve difficult mathematics problems in your mind.

Use *association learning* to help you remember better. Make a link between new facts and some well established old facts. The recalling of old facts will help you remember the new facts.

EXAMPLE 1: When learning the commutative property, remember the word commutative sounds like the word community. A community is made up of different types of people who could be labeled as an "A" group and a "B" group. However, in a community of "A" people and "B" people, it doesn't matter if we count the "A" people first or the "B" people first; we still have the same total number of people in the community. Thus, $a+b = b+a$.

EXAMPLE 2: When learning the distributive law of multiplication over addition, such as $a(b+c)$, remember that distribution is associated with giving out a product. Just remember the distributor is "a" and is giving its products to "b" and "c."

Make up your own associations to remember mathematical properties and laws. Remember, the more ridiculous the association, the more likely you will remember it.

The use of *mnemonic devices* are another way to help you remember. Mnemonic devices are easily remembered words, phrases or rhymes associated with difficult-to-remember principles or facts.

EXAMPLE: Many students become confused when using the Order of Operations. These students mix up the order of the steps in solving a problem, such as dividing instead of first adding the numbers in the parentheses. A mnemonic device to remember the Order of Operations is "Please Excuse My Dear Aunt Sally". The first letter in each of the words represents the math function to be completed from the first to the last. Thus, the Order of Operations are parentheses, exponents, multiplication, division, addition, and subtraction. Now you can make up your own mnemonic devices to remember other mathematics rules.

Acronyms are another memory device to help you learn math. EXAMPLE: FOIL is one of the most common math acronyms. FOIL is used to remember the procedure to multiply two binomials. Each letter in the word FOIL represents a math operation. FOIL stands for First, Outside, Inside, and Last as it applies to multiplying two binomials such as $(2x + 3) (x + 7)$.

To use FOIL multiply the first terms ($(2x) (x)$), the outside terms ($(2x) (7)$), the inside terms ($(3) (x)$), and the last terms ($(3) (7)$). To review the FOIL method, follow this process: $(2x + 3) (x + 7) = (2x) (x) + (2x) (7) + (3) (x) + (3) (7)$ which equals $2x^2 + 17x + 21$. Another way to remember FOIL is to look for the face in **Figure 17**.

Memory output is in the form of verbal or written examinations. Memory output can be affected by test anxiety and test-taking skills. Test anxiety can decrease your ability to recall important information or totally block out information. Your ability to take certain types of tests can also affect your memory output. You might be better at taking multiple-choice tests compared to fill-in-the-blank tests. Some students also have better test-taking skills compared to other students. Techniques to improve your memory output are discussed in the next chapter.

Even though learning disabled students have average to above-average intelligence, they might have more difficulty progressing through the stages of memory. Their sensory input might be reduced due to a visual or auditory processing deficit.

132

Short-term memory may be affected, leaving the student with less ability to recall facts. Timed math tests might measure their learning disability instead of their mathematics knowledge. If you are a learning disabled student, then read Reference I for important information on how learning disabilities affect memory.

FIGURE 17

F (a) (c)
O (a) (d)
I (b) (c)
L (b) (d)

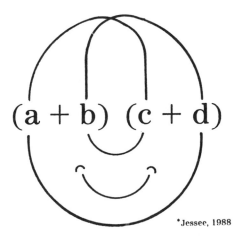

(a + b) (c + d)

*Jessee, 1988

SUMMARY

Learning critical memory techniques begins with understanding the relationship between receiving, storing, and retrieving information. Shifting the information from short-term memory to long-term memory is a crucial process.

While studying, many students don't complete this memory shifting process. Association, recitation, forming mental pictures, and use of mnemonic devices are learning techniques that help transfer math concepts into long-term memory. Use your imagination to adapt these learning techniques to the math material to be understood and learned.

ASSIGNMENT FOR CHAPTER 8

1. Read Reference E — Ten Ways To Improve Your Memory
2. Explain the difference between short- and long-term memory.

3. What are your three best critical memory techniques?

4. Give an example of how you can use association to learn a math principle.

5. Give an example of a mnemonic device that could be used to improve your learning of math principles.

Chapter 9

Improving Mathematics Test-Taking Skills

Comparing Homework Questions to the Actual Test

Most students believe that doing all their homework ensures an "A" or "B" on tests. This is far from true. Doing all the homework and getting the correct answers is very different in many ways from taking tests:

1. While doing homework there is little anxiety. A test situation is just the opposite.

2. You are not under a time restraint while doing your homework; you may have to complete a test in 55 minutes.

3. If you get stuck on a homework problem, your textbook and notes are there to assist you.

4. Once you learn how to do several problems in a homework assignment, the rest are similar. In a test the problems are all mixed up.

5. You have the answers to at least half the problems. Do not develop a false sense of security by believing you can make an "A" or "B" by just doing your homework.

Developing Practice Tests

Developing practice tests is the best way to determine if you are ready for a major test. You can develop a practice test by selecting questions from chapter sections or chapter reviews. Write out some of the questions in random order and wait a day. Then, take the test with the usual time limit. Believe me, you'll locate your problem areas.

Holding a Group Study Session

A better way to prepare for a test is the group method. Hold a group study session several days before the test. Have each student prepare a test with ten questions. On the back of the test have the answers worked out step-by-step. Have each member of the study group exchange his/her test with another member of the group. Once all the tests have been completed, have the author of each test discuss with the group the procedures used to solve those problems.

Make sure the individual or group test is completed at least three days before the real test. Completing these practice math tests will help you increase testing skills. It will also reveal your test problem weaknesses in time to learn how to solve the problem before the real test.

Following General Pre-test Principles

General principles are important when taking any type of test:

1. *Get a good night's sleep before taking a test.* This is true for the ACT, SAT, and even your mathematics tests. If you are going to cram all night and imagine you will perform well on your test with three to four hours of sleep, you're wrong. It would be better to get seven or eight hours sleep and be fresh enough to use your memory in recalling information needed to answer the questions.

2. *Start studying for the test at least three days ahead of time.*

3. *Review only already learned material the night before.*

Winning with the Ten Steps to Better Test-Taking

Once you begin a test, follow the ten steps to better test-taking below:

Step 1: *Upon receiving your test, write down the information that you think you might forget.* Don't put your name on it, do not skim it, just turn it over and write down those facts, figures, and formulas you might not remember during the test. This is called a primary data dump. The data dump provides memory keys for test questions.

EXAMPLE: It might take you a while to remember

how to do a coin word problem. However, if you had immediately turned your test over and written down different ways of solving coin word problems, it would be easier to solve the coin word problems once you had come to it.

Step 2: *Preview the test.* Previewing the test requires you to look through the entire test to find the type of problems and their point value. After previewing, put your name on the test.

Step 3: *Do a second data dump.* The secondary data dump is for writing down material that was jarred from your memory while previewing the test.

Step 4: *Develop a test progress schedule.* When you begin setting up a test schedule, determine the point value for each question. You might have some test questions that are worth more points than other test questions. In some tests, word problems are worth five points and other questions might be worth two or three points. You then have to *decide the best way to get the most points in the least amount of time.* This might mean working the questions worth two to three points first and leaving the more difficult word problems for last.

Step 5: *Answer the easiest problems first and review the answers to see if they make sense.* Start working through the test as fast as you can while being accurate. Answers should be reasonable. For example, the answer to a problem of trying to find the area of a rectangle can't be negative, and the answer to a land rate-distance problem can't be 1000 miles per hour.

Clearly write down each step in order to get partial credit if you end up missing the problem. In most math tests, the easier problems are near the beginning of the first page; you need to answer them efficiently and quickly. This will give you more time for the harder problems.

Step 6: *If you find a problem that you do not know how to work, then automatically skip it.* This could be the type of problem you've never seen before or a problem in which you get stuck on the second or third step. In both cases, skip the problem and go on to the next one.

Step 7: *Review the skipped questions.* When working the

skipped problems, think how you have solved other similar problems as a cue to solving the skipped ones. Also try to remember how the instructor solved that type of problem on the board.

Step 8: *Guess at the remaining problems or do as much work as you can on them.* Do not waste too much time on guessing or trying to work the problems you cannot do.

Step 9: *Review the test for careless errors.* Students usually lose two to five test points on careless errors.

Step 10: *Use all the allowed test time.* Review each problem by substituting the answer back into the equation or doing the opposite function required to answer the question. If you cannot check the problem by the two ways mentioned, then rework the problem on a separate sheet of paper and compare the answers. Do not leave the test room unless you have reviewed each problem three times or until the bell rings. REMEMBER: There is no prize for handing your test in first.

Stapling your scratch paper to the math test, when handing it in, has several advantages:

A. If you miscopied the answer from the scratch paper, you will probably get credit for the answers.

B. If you get the answer incorrect due to a careless error, your work on the scratch paper could give you a few points.

C. If you do get the problem wrong, it will be easier to locate the errors when the instructor reviews the test. This will prevent you from making the same mistakes on the next math test.

Handing in your scratch paper may get you extra points or improve your next test score.

Six Types of Test-Taking Errors

To improve future test scores, you must conduct a test analysis of previous tests. In analyzing your tests, you should look for the following kinds of errors.

(1) misread direction errors
(2) careless errors
(3) concept errors

"SIX TYPES OF TEST-TAKING ERRORS"

(4) application errors
(5) test-taking errors
(6) study errors

Students who conduct math test analyses improve their total test scores.

Misread direction errors occur when you skip directions or misunderstand directions but do the problem anyway.

EXAMPLE 1: You have this type of problem to solve: (x+1) (x+1)

Some students will try to solve for x, but the problem only calls for multiplication. You would solve for x only if you have an equation such as (x+1) (x+1) = 0.

EXAMPLE 2: Another common mistake is not reading the directions before doing several word problems or statistical problems. All too often, when a test is returned, you find only three out of the five problems had to be completed. Even if you did get all five of them correct, it cost you valuable time which could have been used obtaining additional test points.

To avoid misread direction errors, *read all the directions.* If you don't understand them, ask the instructor for clarification.

Careless errors are mistakes made which you would catch automatically upon reviewing the test. Good and poor math students make careless errors. Such errors can cost a student the difference of a letter grade on a test. An infamous careless error in math tests is dropping the sign.

EXAMPLE: $-3x(2x) = 6x^2$, instead of $-6x^2$, which is the correct answer.

Another careless error is not simplifying your answer.

EXAMPLE: Leaving $3x-12/3$ as your answer instead of simplifying it to $x-4$.

Another careless error involves adding fractions:

EXAMPLE: $1/2 +1/3 = 2/5$, instead of $5/6$, which is the correct answer.

When working with students who make careless errors, I ask them two questions. First, "How many points did you lose due to careless errors?" Then I follow with, "How much time was left in the class period when you handed in your test?" Students who lose test points to careless errors are giving away points if they hand in their test papers before the test period ends.

To reduce careless errors you must realize the type of careless errors made and recognize them when reviewing your test. If your major error is not simplifying the answer, then review each answer as if it were a new problem, and try to reduce it.

Concept errors are mistakes made when you don't understand the properties or principles required to work the problem. Concept errors, if not corrected, will follow you from test to test causing loss of test points. Some common concept errors are not knowing:

$(-) (-)x = x$ not $-x$

$-1(2) > x(-1) = 2 < x$, not $2 > x$

$5/0$ is undefined, not 0

$(a+x)/x$ is not reduced to a

the order of operations.

Concept errors must be corrected to improve your next math test score.

Students who have numerous concept test errors will fail the next test and the course — if concepts are not understood. Just going back to rework the concept error problems is not good enough. You must go back to your textbook or notes and learn why you missed those types of problems — not just the one problem itself. You must learn how to work those types of problems in the future. Get help from your instructor now; if you wait, it will be too late.

Application errors occur when you know the concept but cannot apply it to the problem. Application errors usually are found in word problems or by deducing formulas such as the quadratic equation. Even some better students become frustrated with application errors; they understand the material but cannot apply it to the problem.

To reduce application errors, you must predict the type of application problems that will be on the test. Then, you must think through and practice solving those types of problems using the concepts.

EXAMPLE: If you must derive the quadratic formula, then you should practice doing it backwards and forwards, while telling yourself the concept used to move from one step to the next.

Application errors are common with word problems. When doing word problems, look for the key phrases displayed in **Figure 14** to help you set up the problem. After completing the word problem, reread the question to make sure you have applied the answer to the intended question. Application errors can be avoided with appropriate practice and insight.

Test-taking errors apply to the specific way you take tests. Some students consistently make the same types of test-taking errors. Through recognition these bad test-taking habits can be replaced by good test-taking habits. The result will be higher test scores. The following test-taking errors can cause you to lose a lot of points on an exam:

1. *Missing more questions in the 1st-third, 2nd-third, or last-third of a test is also considered a test-taking error.* Missing more questions in the 1st-third of a test could be caused by carelessness when doing easy problems or expe-

riencing test anxiety. Missing questions in the last part of the test could be due to the fact that the last problems are more difficult than the earlier questions, or due to your increasing your test speed to finish the test. If you miss more questions in a certain part of the test consistently, then use your remaining test time to review that section of the test first.

2. *Not completing a problem to its last step is another test-taking error.* If you have this bad habit, then review the last step of the test problem first, before doing an in-depth test review.

3. *Changing test answers from correct ones to incorrect ones is a problem for some students.* Find out if you are a good or bad answer changer by comparing the number of answers changed to correct and incorrect answers. If you are a bad answer changer, then write on your test, "Don't change answers." Only change answers if you can prove to yourself or the instructor that the changed answer is correct.

4. *Getting stuck on one problem and spending too much time on it is another test-taking error.* You need to set a time limit on each problem before moving to the next problem. Working too long on a problem without success will increase your test anxiety and waste valuable time that could be used in solving other problems or in reviewing your test.

5. *Rushing through the easiest part of the test and making careless errors is a common test-taking error for the better student.* If you have the bad habit of getting more points taken off for the easy problems than for the hard problems, review the easy problems first and then review the hard problems.

6. *Miscopying an answer from your scratch work to the test is an uncommon test-taking error, but it does cost some students points.* To avoid these kinds of errors, systematically compare your last problem step on scratch paper with the answer written on the test. In addition, always hand in your scratch work with your test.

7. *Leaving answers blank will get you zero points.* If you look at a problem and cannot figure out how to solve it, do

not leave it blank. Write down some information about the problem or try at least to do the first step. REMEMBER: Writing down the first step of a problem is the key to solving the problem and obtaining partial credit.

8. *The last major test-taking error is not following the ten steps to better test-taking.* Review your test-taking procedures for discrepancies in following the ten steps to better test-taking. Deviating from these proven ten steps will cost you points.

The last type of mistake to look for in test analysis is study error. Study error occurs when you study the wrong type of material or don't spend enough time on pertinent material. Review your test to find out if you missed problems because you did not practice that type of problem or because you did practice it but forgot how to do it during the test. Study errors will take some time to track down. But correcting study errors will help you on future tests.

Sometimes mathematics tests have objective test questions. Objective test questions are in the form of true and false, multiple choice, and matching. To better understand the procedures in taking an objective test, read Reference F — Taking an Objective Test.

Very few mathematics tests have essay questions. However, improving the skills required to take an essay test could give you more time for studying mathematics. Read Reference G — Answering an Essay Test.

SUMMARY

Completing a practice math test several days before the actual exam can locate math areas needing improvement. Correct these math weaknesses by reviewing homework or obtaining help from your instructor or a tutor.

Follow the "Ten Steps To Better Test-Taking" to obtain the most amount of test points in the least amount of time.

Completing a test analysis on your previous math test can help increase test points on the next exam. Test analyses can locate some of your bad test habits. Without test analyses you will probably continue to make the same old test errors and lose valuable test points.

To learn more about mathematics test-taking read Reference H — Studying for Exams and listen to the cassette tape, *How to Ace Tests* (available from your college bookstore or by mail order by using the order form in the back of this book).

If you have a learning disability, then your test-taking skills and test accommodations are especially important. You may want to use the Strategy Cards For Higher Grades as an auxiliary learning aid. Test accommodations could come in the form of extended time or large-print tests. Read Reference I to learn more about test accommodations for learning disabled students.

ASSIGNMENT FOR CHAPTER 9

1. Read Reference F — Taking an Objective Test.
2. Read Reference G — Answering an Essay Test.
3. Read Reference H — Studying For Exams.
4. Listen to the cassette tape *How to Ace Tests*.
5. How can you correctly do all the homework and still not score high on math tests?

6. How can you use a study group to prepare for a math test?

7. What are the ten steps to better test-taking?

8. What did you learn from your test analyses?

9. How can you improve your math test-taking skills?

NOTES

Chapter 10

Taking Control Over Mathematics

Developing an Internal Locus of Control

The way you can take control over mathematics is by developing an internal locus of control, avoiding learned helplessness, and decreasing procrastination.

Locus of control is *where* a student places the control over his life; in other words, who or what the student feels controls his behavior and grades. Students who feel that conditions beyond their control prevent them from getting good grades have an external locus of control. These students blame instructors, home conditions, and money problems for their poor grades and feel they can do nothing about their problems. In essence, external students feel their lives are controlled by outside forces such as fate or the power of other people.

Other students feel they have the power to control their situation and this power comes from within. These *internal students take responsibility for their success, while external students reject responsibility.* Internal students believe that they can overcome most situations, since results depend on their behavior or personal characteristics. Internal students accept the responsibility for their behavior and realize that studying today will help them pass the mathematics test scheduled for next week. Internal students can delay immediate rewards; they study tonight for a test tomorrow instead of going to a party.

In general, locus of control means that students who are internal will work harder to meet their educational goals than will external students. The internal student can relate today's behavior (e.g., studying, textbook reading) to obtaining a college degree and gainful employment. The external student, on the other hand, cannot connect the behavior of studying today with obtaining adequate grades and future career opportunities. Thus, internals are more oriented towards making high mathematics grades than externals.

Externals can change into internals by taking more responsi-bility for their lives and for obtaining their education. You can take more responsibility by developing and accomplishing short-term goals and long-term goals.

Short-term goals are goals developed and accomplished within a day or week. For instance, a goal of studying math today between 7:00 pm and 9:00 pm. A long-term goal, for example, could be earning an "A" or "B" in the math course for the semester.

The steps to obtaining your short-term goals or long-term goals must be thought out or written down. Rewarding yourself after meeting short-term goals increases your internal control by strengthening the connection between your behavior and the obtained reward. Short-term goal successes lead to greater suc-cesses in obtaining long-term goals.

Avoiding Learned Helplessness

As students become more external they develop learned help-lessness. Learned helplessness means believing that other peo-ple or influences from instructors, poverty or "the system" con-trols what happens to them. These students adopt the attitude of "Why try?" Students who have failed mathematics several times may develop learned helplessness.

Total lack of motivation to complete math assignments is a good example of learned helplessness. Students who develop learned helplessness can break this bad habit. In the past, a student may have completed the math assignments but was not successful in obtaining the grade he wished in the course. This led to the attitude of "Why try", because he tried several times in the past to be successful in math and still failed the course.

The problem with this thinking is the way he actually tried to pass the math course. His learning processes, anxiety reduction, and test-taking techniques were not as effective as the ones discussed in this book. It was like trying to remove a flat tire with a pair of pliers instead of using a tire iron.

Now that you have the "tire iron" (*Winning at Math*), the question is, "Are you motivated enough to put forth the effort to use the "tire iron" and make a good grade?" Take responsibility, and you are on your way to winning at math.

Overcoming Procrastination, Fear of Failure and Fear of Success

Procrastination is an inefficient way to take control over mathematics. Students may procrastinate by not reading the textbook or not doing their homework due to fear of failure, fear of success, or rebellion against authority. Some students who fear failure procrastinate to avoid any real assessment of their true ability. By waiting too long to begin work on a paper or studying for a test, real ability is never measured. Thus, you can never learn the degree of "goodness" or "badness" of your academic ability.

Other students who have fear of failure could be "closet" perfectionists. Perfectionists usually expect more of themselves than can be possibly obtained.

FOR EXAMPLE: A math student was failing the course at midterm decides to drop it, retake it next semester, and set a goal to make an "A". After making a "C" on the first major test he became frustrated, started procrastinating in his math studies, and expected to fail the course. This student believed he'd rather not even try to pass the course if he couldn't make an "A."

Being a perfectionist isn't related to how high you set the goal but the unrealistic nature of the goal itself.

Fear of success means not putting an all-out effort towards becoming successful. Some students believe becoming too successful will lose them friends, lovers, or spouses, or will result in guilt for being more successful than family or close friends.

This "fear of success" can be generalized as "fear of competition" in making good grades. These students don't fear the chance of making low grades when competing. They fear they will not be liked by others if they make high grades.

For example, a math student may fear that by studying too much he will make the highest test grade and set the grading curve. He has more fear that students will not like him due to his high grades than the fear of just making average grades. These students need to take pride in their learning ability and let the other students take responsibility for their own grades.

"REBELLING AGAINST AUTHORITY"

Rebelling Against Authority

The third cause of procrastination is the desire to rebel
against authority. Students believe that by handing in their

homework late or by missing the test they can get back at the instructor.

These students usually lack self-esteem and would rather blame the instructor for their poor grades than take responsibility for completing their homework on time. Rebelling against the instructor gives them a false sense of control over their lives. However, the rebelling students are fulfilling the exact expectations placed on them by their instructors: becoming academic failures. These students discover, often too late, they are only hurting themselves.

Procrastination is not a simple issue. Students procrastinate for various reasons. Procrastination, though, is mainly a defense mechanism that protects self-esteem. Most students who procrastinate have poor mathematics grades.

SUMMARY

Taking control over mathematics means becoming more internal, avoiding learned helplessness and decreasing procrastination. Effectively using all of the 25% of your affective learning characteristics is required to significantly improve math performance.

You can start becoming internal by taking the responsibility for following the suggestions in Chapters Five through Nine while setting and accomplishing realistic short-term academic goals. You can avoid learned helplessness by not giving up on making a good grade in mathematics, and if needed, getting help from your instructor and counselor. You can start decreasing procrastination by following the suggestions in Chapter Three (Learning How to Reduce Mathematics Anxiety), Chapter Four (Developing Effective Study Management Skills), and limiting its use as a defense mechanism.

Remember, students who have followed the suggestions in this book have significantly improved their mathematics grades as compared to similar students who did not use these suggestions.

ASSIGNMENT FOR CHAPTER 10

1. Meet with your mathematics instructor at least three times during the semester to get feedback on your course progress.

2. Meet with your instructor for this course at least once to discuss specific procedures to improve your mathematics grades.

3. In the space below write an analysis of your reasons for procrastination in mathematics and how you are going to overcome them.

GLOSSARY

ACRONYM — a memory technique in which one or more words are made up of the first letter of each of the words comprising the information you wish to remember. For example, ROY G BIV is an acroynm representing the colors of the rainbow — red, orange, yellow, green, blue, indigo and violet.

ADJUNCT MATH FACULTY — part-time math instructors who are employed to teach one to three courses. They usually do not have office hours and usually cannot meet with students needing extra help.

AFFECTIVE CHARACTERISTICS — characteristics students possess that effect their course grades, excluding cognitive entry skills. Some of these characteristics are anxiety, study habits, study attitudes, self concept, motivation, and test-taking skills.

ASSOCIATION LEARNING — a memory technique used to relate new information to be learned to old information you already know.

COGNITIVE ENTRY SKILLS (MATH) — math knowledge a student possesses when first beginning a particular math course.

CONDITIONED RESPONSE — a habit developed by doing the same behavior over and over again.

CUE-CONTROLLED RELAXATION — a relaxation response technique in which a student can relax himself by repeating certain cue words to himself. A good example of this is that upon hearing certain old songs (cue words) your feelings (emotions) often change.

DISCUSSION OF RULES — part of the modified two-column note-taking system in which students write down their lecture notes and important rules used to solve the problems present - ed in class.

DISTRIBUTIVE LEARNING — a learning system in which you spread your homework on a particular subject over several days instead of trying to do it all at one time. For example, studying for about an hour and taking a five to ten minute break before continuing to study.

EFFECTIVE LISTENING — a behavior in which you sit in the least distractive area of the classroom and become actively involved in the lecture.

EXTERNAL STUDENT — a student who believes that he/she is not in control of his/her own life and that he/she cannot obtain his/her desired goals, like making a good grade in math. External students blame their teachers, parents — anyone and anything except themselves, for their failures.

FEAR OF FAILURE — a personal defense mechanism by which a students puts off doing his/her homework so he/she may have an excuse when he/she does poorly in or fails the course. Thus the student's real ability is never measured.

FEAR OF SUCCESS — a personal defense mechanism by which a student doesn't put forth all his/her effort to obtain good grades. Many students believe that by becoming too successful they will lose friends or may be expected to make good grades all the time.

GLOSSARY OF TERMS — a section in the back of a notebook developed by the student which contains a list of key words or concepts and their meanings.

HIGHLIGHTING — underlining important material in the textbook or in your notes with a magic marker or felt-tip pen.

INTERNAL STUDENT — a student who believes that he/she is in control of his/her own life and can obtain his/her desired goals — like making a good grade in math.

LEARNED HELPLESSNESS — a lack of motivation due to repeated tries to obtain a goal (like passing math) but failing to obtain that goal. An attitude of "Why try?" develops because of numerous previous failures.

LINEAR LEARNING PATTERN — a learning pattern in which one concept builds on the next concept. The ability to learn new math concepts is based on your previous math knowledge. Not knowing underlying math concepts causes gaps in learning, which often results in lower future test scores and even failure.

LOCUS OF CONTROL — the belief that one is in control of his/her own life, or that other people or events are controlling his/her life.

LONG-TERM MEMORY — the last part of the memory chain which retains unlimited information for long periods of time and is considered to be a person's total knowledge.

MASS LEARNING — bunching-up all your learning periods at once. For example, trying to complete all your math homework for the last two weeks in one night. This technique is an ineffective way of learning math.

MATHEMATICS ACHIEVEMENT CHARACTERISTICS — characteristics that students possess which affect their grades, such as previous math knowledge, level of test anxiety, study habits, study attitudes, motivation, and test-taking skills.

MEMORY — the process of receiving information through your senses, storing the information in your mind, and recalling the information for later use.

MENTAL PICTURE — a memory technique in which you visualize the information you wish to learn by closing your eyes and forming an image of the material in your mind's eye.

MNEMONIC DEVICE — a memory technique in which you

develop easy-to-remember words, phrases, and rhymes, and relate them to difficult-to-remember concepts.

NOTE CARDS — 3" X 5" index cards; students write important concepts on the front of the card and an explanation of the concepts on the back of the card.

NOTE-TAKING CUES — signals given by instructors to their classes which indicate that the material they are presenting is important enough that the students may be tested on it. Notes should be taken on this material.

PERFECTIONIST — one who expects to be perfect at everything he does, including making an "A" in math when it may be, for him, virtually impossible.

PROCRASTINATION — a personal defense mechanism in which one puts off doing certain tasks, like homework, in order to protect one's self-esteem.

QUALITY OF INSTRUCTION — the effectiveness of math instructors when presenting material to students in the classroom and math lab. This effectiveness depends on the course textbook, class atmosphere, teaching style, extra teaching aids (videos, audio tapes), and other assistance.

RELAXATION RESPONSE — a learned technique which decreases emotional anxiety and/or disruptive thought patterns, allowing you to think more clearly.

RE-WORKING NOTES — the process of reviewing class notes to re-write illegible words, fill in the gaps, and add key words or ideas.

SENSORY REGISTER — the first part of the memory chain that receives the information through your senses (seeing, hearing, feeling, and touching).

SHORT-TERM MEMORY — the second part of the memory chain which allows you to remember facts for immediate use.

These facts are soon forgotten.

SKIMMING — the first step in reading a textbook. It involves over-viewing the chapter to get a general understanding of the material.

STUDY BUDDY — a student who is usually taking the same math course as yourself whom you can call for help when you have difficulty doing your math homework.

TEST ANXIETY — a learned emotional response or thought pattern response that disrupts or delays a student's ability to recall information needed to solve the problems.

TEST ANALYSIS — a process of reviewing previous tests for consistent misread directions, careless concept, application, test-taking and study errors in order to help prevent their future occurrence.

TIME MANAGEMENT — a process of gaining control over time in order to help you obtain your desired goals. Using a study schedule is an example of gaining control over time.

TOOLS OF YOUR TRADE — any materials you require to begin studying.

WEEKLY STUDY GOALS — the amount of time scheduled for studying each of your subjects over the period of a week.

REFERENCES

REFERENCE A

MATH POST-TEST
EVALUATION

Read each of the items below. Choose the statement in each group which is true of you. Indicate what you *actually do* rather than what you *should do* by circling a, b, or c. BE HONEST.

1. I:
   a. seldom study math every school day.
   b. often study math every school day.
   c. almost always study math every school day.

2. When I register for a math course, I:
   a. seldom select the best math teacher.
   b. often select the best math teacher.
   c. almost always select the best math teacher.

3. I:
   a. seldom become anxious and forget important concepts during a math test.
   b. often become anxious and forget important concepts during a math test.
   c. almost always become anxious and forget important concepts during a math test.

4. I:
   a. seldom study math at least 8 to 12 hours a week.
   b. often study math at least 8 to 12 hours a week.
   c. almost always study math at least 8 to 12 hours a week.

5. Each week, I:
   a. seldom plan the best time to study math.
   b. often plan the best time to study math.
   c. almost always plan the best time to study math.

6. I:
   a. seldom use an abbreviation system when taking notes.
   b. often use an abbreviation system when taking notes.
   c. almost always use an abbreviation system when taking notes.

7. When I take math notes, I:
   a. seldom copy all the steps to a problem.
   b. often copy all the steps to a problem.
   c. almost always copy all the steps to a problem.

8. When I become confused in math class, I:
   a. seldom stop taking notes.
   b. often stop taking notes.
   c. almost always stop taking notes.

9. I:
   a. seldom fail to ask questions in math class.
   b. often fail to ask questions in math class.
   c. almost always fail to ask questions in math class.

10. I:
    a. seldom stop reading the math textbook when I get stuck.
    b. often stop reading the math textbook when I get stuck.
    c. almost always stop reading the math textbook when I get stuck.

11. When I have difficulty understanding the math topic, I:
    a. seldom go to the instructor or tutor.
    b. often go to the instructor or tutor.
    c. almost always go to the instructor or tutor.

12. I:
    a. seldom review class notes or read the textbook assignment before doing my homework.
    b. often review class notes or read the textbook assignment before doing my homework.
    c. almost always review class notes or read the textbook assignment before doing my homework.

13. I:
    a. seldom fall behind in completing math homework assignments.
    b. often fall behind in completing math homework assignments.
    c. almost always fall behind in completing math homework assignments.

14. After reading the math textbook, I:
    a. seldom mentally review what I have read.
    b. often mentally review what I have read.
    c. almost always mentally review what I have read.

15. There:
    a. seldom are distractions that bother me when I study.
    b. often are distractions that bother me when I study.
    c. almost always are distractions that bother me when I study.

16. I:
    a. seldom do most of my studying the night before the test.
    b. often do most of my studying the night before the test.
    c. almost always do most of my studying the night before the test.

17. I:
    a. seldom develop memory techniques to remember math concepts.
    b. often develop memory techniques to remember math concepts.
    c. almost always develop memory techniques to remember math concepts.

18. When taking a math test, I:
    a. seldom start on the first problem and work the remaining problems in their numbered order.
    b. often start on the first problem and work the remaining problems in their numbered order.
    c. almost always start on the first problem and work the remaining problems in their numbered order.

19. Even when time permits, I:
    a. seldom check over my test answers.
    b. often check over my test answers.
    c. almost always check over my test answers.

20. When my math test is returned, I:
    a. seldom analyze the test errors.
    b. often analyze the test errors.
    c. almost always analyze the test errors.

MATH POST-TEST SCORING FOR STUDY SKILLS

Put the correct amount of points for each item in Section A and Section B to obtain your score. The order of the items are different for Sections A and B.

SECTION A   POINT VALUE FOR EACH QUESTION

Items	Seldom (5 points)	Often (3 points)	Almost Always (1 point)
1.	_____	_____	_____
2.	_____	_____	_____
4.	_____	_____	_____
6.	_____	_____	_____
9.	_____	_____	_____
11.	_____	_____	_____
13.	_____	_____	_____
14.	_____	_____	_____
15.	_____	_____	_____
17.	_____	_____	_____
20.	_____	_____	_____
TOTAL	_____ +	_____ +	_____ = _____

SECTION B   POINT VALUE FOR EACH QUESTION

Items	Seldom (1 point)	Often (3 points)	Almost Always (5 points)
3.	_____	_____	_____
5.	_____	_____	_____
7.	_____	_____	_____
8.	_____	_____	_____
10.	_____	_____	_____
12.	_____	_____	_____
16.	_____	_____	_____

	Seldom	Often	Almost Always
18.	_____	_____	_____
19.	_____	_____	_____
TOTAL	_____ +	_____ +	_____ = _____

_____ + _____ = _____

SECTION A   SECTION B   GRAND TOTAL

A score of 70 or below means that you have poor math study skills and need to read this book again to obtain its valuable information.

A score between 70 and 90 means that you have good math study skills, but you can improve by reviewing certain sections of this book.

A score above 90 means that you have excellent math study skills.

REFERENCE B

STRESS

There are many definitions of personal stress. Webster's Dictionary describes stress as being "Tense, strained exertion: As the stress of war affects many people." Kenneth Lamott, in his book, *Escape from Stress*, defines this condition not only as the well-publicized "executive stress" but, the results of a significant disruption in one's environment which can lead to a stress response syndrome. Stress can also be induced by chronic minor frustrations such as being late to class or a day late paying bills. Psychiatrist Ainsley Mears describes stress as a disease which creates other diseases. Indeed, stress is beginning to be looked on more as a health problem in today's anxiety-producing world.

Stress, no matter how one defines it, will lead to chronic or acute physical and psychological disabilities. Stress leads to distress with the results of mental and/or bodily suffering. Some mental stress problems can be explained by the fight-or-flight response, which was established as a survival mechanism in the early days of mankind. When our ancestors were faced with a threat, hormones automatically were released into the bloodstream and the autonomic nervous system prepared the body for immediate action. This action resulted in a fight or flight from the situation. During the fight or flight, the body metabolized the "stress hormones" and at the end of the ordeal, the person was exhausted but not under stress.

The fight-or-flight response is still active and essential in today's life to survive in warfare, dangerous sports and everyday driving. Yet we have a problem; there are numerous stress events in our lives in which we cannot activate the fight or flight response. A supervisor may ask if we have finished that report or our spouse may ask when we will be home. As we face these stress situations, our autonomic nervous system, just as it did thousands of years ago, releases the "stress hormones" which increase our blood pressure, heart rate, rate of breathing and metabolism, preparing us for the event. But now the event is one which is only socially undesirable instead of an act of survival. Unfortunately we are still left with our increased body functions.

If this process is repeated over a period of time, one would become anxious, uptight, or short-tempered, leading to mental stress and physical problems. The physical problems manifest themselves as high blood pressure, peptic ulcers, and various heart diseases, which in turn shorten life. In short, uncontrolled stress will lead to a shorter, more miserable life.

One would now ask, "What can I do to control the fight-or-flight response?" The answer is nothing, since it is an autonomic system, but you can control the effects of the response. As mentioned above, one side effect of the fight-or-flight response is tension. Sports of all kinds are frequently used to combat stress. So are music, painting, and other hobbies. A trip out to the country or to the beach will relieve the pressures of the day. Alcohol and other drugs such as tranquilizers, barbiturates, and marijuana will also reduce tension but might have dangerous side effects.

There are other ways to reduce tension not mentioned above, but most of them fall into two areas:

(1) chemical

(2) situational.

However, there is a third alternative which Dr. Benson calls "an innate protective mechanism against "overstress," which will counter the effects of the fight-or-flight response." What Dr. Benson is referring to is the relaxation response. The relaxation response can be induced by various exercises or rituals that have been used over generations. Some examples are: Zen, Yoga, T.M., self-hypnosis, biofeedback, progressive relaxation, and autogenic training. The last two examples mentioned above are taught in the audio cassette *How to Reduce Test Anxiety*.

Relaxation Response

When anxiety or tension is brought on by a stressful event, the quality of feelings may be different depending upon the event. Also, some people will experience this anxiety in different ways such as a feeling a "knot" in their stomach or their "mind racing in circles." The person who has a tense stomach is experiencing somatic anxiety and the other person is experiencing cognitive anxiety. However, there can be variations of both types of anxiety in one person, although one is usually dominant.

Given that these are the two basic forms of anxiety, one type of relaxation response should have more effect on cognitive anxiety and another a greater effect on somatic anxiety. Research shows cognitive and somatic anxiety can be reduced if the proper relaxation response is used.

In general, progressive relaxation reduces somatic anxiety and the Benson Method reduces cognitive anxiety. Although autogenic training (a passive procedure to attain relaxation) seems mostly to effect somatic anxiety, it also effects cognitive anxiety. Further explanations of each relaxation response will follow.

Progressive relaxation was developed by Dr. E. Jacobson of the University of Chicago in 1938 and is probably the most widely used relaxation technique today. This relaxation technique entails systematically tensing and relaxing about sixteen different muscle groups. The therapist instructs the subject that on his command the subject will tense one muscle group and then relax it. Upon relaxing, the subject will focus his attention on the muscle group. This sequence of tensing and relaxing one muscle group will be systematically applied to each major muscle group of the body. The results will be somatic relaxation.

Autogenic training was first developed by J. H. Schultz in the 1900's. This procedure is a passive way to achieve somatic relaxation. The subject is placed in a comfortable position such as lying on a mat or in a reclining chair. Verbal instructions are given to relax different parts of the body. Suggestions of heaviness and warmth are given to the subject to induce the feeling of relaxation. Once all parts of the body are relaxed the therapist has the subject imagine tranquil scenes. The typical scenes are of beaches, the forest, and listening to music. This positive imagery helps the subject go deeper into relaxation and reduces cognitive anxiety. The results are both somatic and cognitive reduction in anxiety.

Daily Use of Relaxation

Sometimes during the day we feel ourselves getting tense and tired. The next thing someone says to us might cause us to lose control. There is not enough time to use progressive relaxation. What do we do?

The answer is twofold:

(1) First, we do a short relaxation response which might be palming.

(2) Second, we initiate differential relaxation as one of several preventive measures.

Palming can be done while one is seated with elbows resting on a table or desk. The eyes are closed and covered with the palms of the hands. To avoid exerting any pressure upon the eyeballs, the lower part of the palms should rest upon the cheek bones and the fingers upon the forehead. Once the light is blocked out one can start visualizing some relaxing scenes developed during autogenic training. This should reduce one's anxiety to a coping level.

As a preventive method, differential relaxation should be practiced. Differential relaxation is a way to relax while being active. It involves learning to differentiate between those muscles necessary to perform a task and those which are unnecessary. The muscles that are not necessary to perform a task should be relaxed to conserve nervous and muscle energy. For instance, while sitting and reading a book, you can relax your legs while only maintaining minimal tension in the upper part of your body. As you go through your daily activities, be aware of needless tension in some parts of your body, and relax those body parts.

The techniques mentioned above cover only two ways to control tension in our daily lives. The best way to control tension is to be prepared for it by practicing the relaxation responses best suited to your life style. Prevention is the best policy in coping with stress.

REFERENCE C

TEN STEPS TO IMPROVING
YOUR STUDY SKILLS

Improving your study skills can be a great educational equalizer. By following these ten steps your grades can improve. Educators have proven that effective studying is the best way a student can improve grades. However, most students have not been taught how to study while attending elementary school, middle school or high school. Most teachers, for some reason, think students are born with the ability to study. Students, on the other hand, look to their teachers for assistance in learning study skills only to find that many teachers don't know how to effectively teach them.

Students now have the opportunity to learn the difference between good and poor study skills. Educational research has demonstrated the most effective note-taking, reading, memorizing, and reviewing skills required to make good grades. The following ten steps have been proven by research, and, when used correctly, will improve study skills.

1. *Don't study for more than one hour at a time without taking a break.* If, in fact, you are dong straight memorizing, don't spend more than 25 to 35 minutes. Difficult material should be learned in short study periods and reviewed more often. After studying for 45 minutes to an hour your ability to retain material dramatically decreases. If you have been studying for two or three hours without taking a break, you probably didn't retain the information.

Educators say you learn best in short study sessions. Spreading your studying time over several days is more effective than studying four hours in a row on the same subject. This is especially true when studying math. More time is required to assimilate the concepts through practice between lecture periods. Spreading out your study periods over several days will increase your retention of that material.

2. *Condition yourself to a study area.* Use the association learning concept. When you first start studying, try to study the

same subject in the same place at the same time. Select a quiet place at school and in your home to study. After a while you will discover when you get to that time and place you will be already thinking about that subject. Training your brain to "think math" at specific times will decrease warm-up time. You will also save emotional time and energy once required to psych yourself up for learning math. You will remember more math. After studying, reward yourself by doing something you want to do (watch TV, go to a party). Experts know that positive reinforcement of a behavior (studying) will increase that behavior.

When beginning a study session, start with your most difficult subject first and work towards the easiest subject. Your most difficult subject will take more concentration and effort which requires a fresher and alert mind. Less effort is required to learn easier material later on as your mind becomes tired.

3. *Don't study when you are tired.* Educators know that students have different times during the day that are least productive for studying. Don't study during that time. Discover the times you study best: morning, afternoon, or evening. Then schedule most of your study time for that part of the day or night.

4. *Separate the studying of subjects that are alike.* Your brain has a right and left side. Studying similar subjects one after the other will tire out the same side of the brain. This will decrease your study efficiency. Study math for an hour followed by English or history, not accounting or computers.

5. *Study for your classes at the best time.* You will forget most of the material learned in class right after leaving the room. *To best retain the lecture material, rework class notes or start doing the homework as soon as possible after class.* Prepare questions for the next lecture period. Waiting two to three days to learn the same material will cause confusion and more study time.

6. *Use the best note-taking system for you.* For each subject, use an 8 1/2" by 11" spiral notebook with pockets to store handouts and old tests. Make sure your notebook has enough paper for the entire semester.

When your instructor lectures from the textbook, use the three-column system. (**Figure 18**). Make a column two inches down the left-hand side of the note page for writing key words. The middle column should be made three inches wide and used for lecture notes. The right side should also be three inches wide and be used for textbook notes. Put a two-inch space across the bottom of the page for a summary of lecture and textbook notes.

Leave several pages in the back of the notebook for textbook notes. Label the first page "Chapter One" and make a list of key words or concepts you didn't understand while reading the chapter. Number and write a brief explanation of each concept or key word. Use the same procedure for each chapter. Review these notes as often as possible.

7. *Read and study the text at the same time.* How many times have you finished a reading assignment and could not remember what you just read? Researchers have discovered that it is more efficient to read and study at the same time. This procedure will take a little longer, but you will remember more of the material.

8. *Make up a symbol system for textbook and notes.* When reading your textbook, make up a symbol system to mark important materials for further study. The symbols can be boxes, stars, question marks, circles and check marks. Each symbol represents a different message, such as: Very important material — star, don't understand material — question mark, and repeated material from textbook or notes — check mark. Developing your own symbol system will help you learn the important material.

9. *Don't buy underlined textbooks.* Buy new textbooks or used textbooks that are not underlined. Underlined textbooks represent information another person thinks is important, not what you think is important. If you have to buy an underlined textbook, then use a different color to underline the text.

10. *Use your full ability to memorize.* Don't memorize material by reading it over and over; this is the least efficient way. Use as many of your senses as possible when memorizing material. When memorizing concepts:

> ★ Say the words out loud so you can hear them
> ★ Record important concepts on cassette tape and play them back while you are driving or relaxing
> ★ Visualize the important concepts by closing your eyes and imagining those concepts in your mind
> ★ Write down major concepts several times and say them to yourself

FIGURE 18

THREE—COLUMN SYSTEM

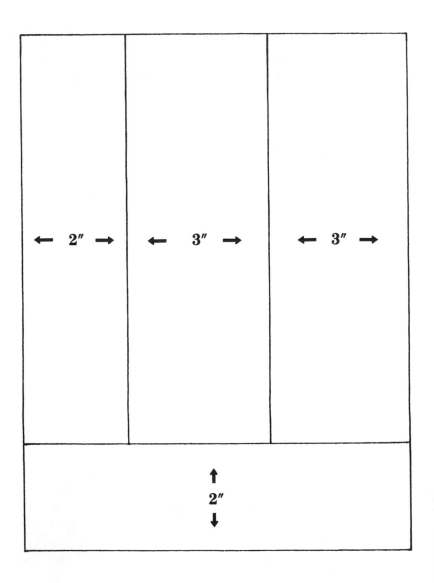

★ Use association learning to tie in learning new material with information you already know. Relate personal facts such as your birthday with facts to be learned.

★ Use acronyms such as ROY G BIV, which is the key to remembering the colors of the rainbow (red, orange, yellow, green, blue, indigo, violet).

★ Use mnemonics, which are phrases to help remember the order of different concepts. For example, the phrase "Please Excuse My Dear Aunt Sally" represents the Order of Operations to complete a math problem (parentheses, exponents, multiplication, division, addition, subtraction).

All of the above steps must be followed to increase the chances of you improving your grades. Just using one or two of the above steps will not have enough impact on your learning to improve grades. Use all the steps and watch your grades improve.

REFERENCE D

SUGGESTIONS TO TEACHERS FOR IMPROVING STUDENT MATH STUDY SKILLS

1. Show a film on test anxiety to students.
2. Help students learn to relax before tests are given.
3. Teach students how to take good notes in math class.
4. Emphasize that students should spend more time on math homework.
5. Encourage students to complete their most difficult homework assignments first. Usually, this means math homework.
6. Encourage students to read ahead in the math textbook in order to prepare questions for the instructor and make an informal outline.
7. Prepare a list of math vocabulary words for the students.
8. Inform students about the audio tapes on different math topics located in the math lab.
9. Encourage students to develop study schedules and set up group study times.
10. Encourage students to construct fake tests and time themselves while taking them (prior to taking genuine tests).
11. Aid in developing the students' self-esteem so they will think about succeeding in the class rather than fearing failure.
12. Suggest that students do all the example problems in the text for practice.
13. Encourage students to write down questions for the instructor while doing homework.
14. Dispense resources for tutor assistance and encourage the use of such assistance.
15. Make students aware of the time allotted them while taking a math test so that they may plan their work.
16. Discuss the importance of class attendance with your students.
17. Advise students to schedule their classes to leave room for study time.
18. Develop a math hot-line for students to call the Math Lab.
19. Videotape different math classes so that students can review important math topics.

20. Encourage students to verbalize (silently) problems the instructor writes on the board. They could then solve the problems or silently verbalize the methods of solution.

21. Encourage students to interview a particular instructor before actually signing up for the class.

22. Suggest that students make note cards to remind them of how to solve various math problems.

23. Advise students to get help early in the semester, should they need assistance, before they get too lost in the course.

24. Suggest that students recite back the materials they have read in the math textbook for understanding.

25. Suggest that students take notes while they are doing problems.

26. Tell students to copy all the information that is put on the board.

27. Recommend the book, *How To Solve Word Problems: A Solved Problem Approach* by Mildred Johnson for use by students having difficulty with word problems.

28. Advise students to prepare for math tests through practice rather than memorizing.

29. Encourage students to do math homework every day.

30. Emphatically discourage missing math classes.

31. Allow students to call their instructors for permission to attend the same course that is taught at a different time.

32. Hold students responsible for materials covered in classes they have missed.

33. Develop a dictionary of math vocabulary words.

REFERENCE E

TEN WAYS TO IMPROVE YOUR MEMORY

1. *Space practice* — Spacing study periods is more efficient than learning material all at once. Long study periods do not allow an opportunity to consolidate what you have learned.

Time and spacing varies: four 1-hour periods result in better recall of material than one 4-hour session.

In the evening, you should study in short, spaced periods. You should then go to bed and get up the next morning and review what you learned the night before.

2. *Active recitation* — As you read and learn information, stop often and repeat to yourself what you learned. Put into your own words what you have just learned. This focuses your attention on the material at hand and gives you repeated practice in retrieving information that has been stored. It will help you to recall (remembering on your own) rather than just recognizing the information stored.

If it helps you, write the facts down in outline form at the same time that you are reciting out loud the information.

Sometimes it helps to study with someone else. They can ask you questions about the information that you have just read or studied. Repeat out loud to them the important facts learned.

3. *Overlearn the material* — Continue to practice beyond the point of bare mastery. Even though you are able to recall it once, continue to practice over and over again. This will increase the amount of material that you will be able to remember later.

4. *Recall* — recounting of something learned.

Recognition — identification of something learned.

Relearning - learning material that was learned previously.

Relearning is reviewing material learned earlier in the school term in order to refresh your memory, rather than cramming all at one time. The more you review during the school term, the less time you will have to spend going over old material for the exam or evaluation.

5. *Use physical resources to aid memory* — This means having other devices to remind you when to complete tasks. These memory devices can be notes, appointment book, calendar, lists,

and people. Intelligent people use these memory devices to aid in the recall of information.

For example, to remember to do something when you get home, write a note to yourself and put it in your pocket. When you empty out your pockets the note will remind you complete that task.

Another physical resource to aid memory is the "to do" list. This is a list of items to be completed during the day. Check off the items as you complete them and add more items to the list as needed.

6. *Use chunking* — This is a technique in which several items are grouped together in short-term memory and then are rehearsed together as one block or chunk. For example, the acronym ROY G BIV is remembered as one chunk of information instead of the colors of the rainbow (red, orange, yellow, green, blue, indigo, violet). Each letter of ROY G BIV represents the first letter of of each color of the rainbow.

7. *Use mediation* — This means to understand the material by making real words, connections between words, or sentences out of the material. For example, please excuse my dear aunt sally is a mediating device to remember the order of operations in solving a math problem. The first letter of each word represents the first letter of each order of operations (parenthesos, exponents, multiply, divide, add, subtract).

8. *Use associations* — This means connecting what you want to remember with something you already know. For example, you want to remember that lysomes are part of a cell that kills bacteria. You then associate the word lysomes with a common cleaner called lysol that also kills bacteria.

9. *Mental Imagery* — This is used to form mental pictures of items or events you are trying to remember. To help recall an item or event, picture the place in which you learned it, imagine the surroundings, feelings and even the place you sat. Also, to remember items or events close your eyes and mentally picture the information in your mind's eye. Use mental imagery to remember and recall items or events.

10. *Use a peg word list* — This procedure involves learning a series of names for concrete items that you can remember in correct order. Develop a symbol for each item and associate the symbol with a peg word. For example, memorize these rhyming

peg words: one — gun, two — shoe; three — tree; four — door; five — hive; six — sticks; seven — heaven; eight — gate; nine — line; and ten — hen.

Your next step is to make up another list of five to ten words and numbers using your imagination to associate the number and word. Use this system to memorize a list of words and see how easy it can be accomplished.

REFERENCE F

TAKING AN OBJECTIVE TEST

Multiple Choice

True-False

Matching

1. Before you start answering any questions, preview the entire test. Survey the test to discover how many questions there are and of what type. Set a time limit so that you will have at least five minutes at the end to recheck your test.

2. Read the directions carefully, making sure you understand exactly what is expected.

3. Find out if you are penalized for guessing. If not, always guess and don't leave any unanswered questions.

4. Read each question carefully, underlining key words.

5. Anticipate the answer and then look for it. Read all the alternatives before answering.

6. Don't read into questions what is not there.

7. When your anticipated answer is not one of the options, discard it and concentrate on the given ones systematically.

8. When two or more options look correct, compare them with each other. Study them to find what makes them different. Choose the more encompassing option unless the question requires a specific answer.

9. Pass over the difficult or debatable questions on your first reading and then come back to them after completing those of which you were sure.

10. Use information from other questions.

11. In all questions, especially the true-false type, look for specific determiners. Words such as rarely, usually, sometimes, and seldom allow for exceptions; never, always, no, and all indicate no exceptions.

12. Mark statements true only if they are true without exception. If any part of the statement is false, the whole statement is marked false.

13. Stay in one column of a matching test, usually the column with the definition and work backwards to find the word or symbol that matches it. Be sure to find out if the answers can be used more than once.

14. If you know you made an error, change your first answer. If it is just a guess, keep your first impression.

REFERENCE G

ANSWERING AN ESSAY TEST

Short or Long Answer
Fill-in-the Blank
Sentence Completion

1. Make a brief survey of the entire test. Read every question and the directions. Plan to answer the least difficult questions first, saving the most difficult for last.

2. Set a time schedule and periodically check your progress to maintain proper speed. With six questions to answer in sixty minutes you should allow a maximum of ten minutes per question. If your ten minutes pass and you have not finished the question, continue to the next one and go back to it later. Do not sacrifice any question for another.

3. Read the question carefully. Underline key words: list, compare, WWII, political and social, art or music, etc. As you read, jot down the points that occur to you beside that question.

4. Organize a brief outline of the main ideas you want to present. Place a check mark alongside each major idea and then number them in order of presentation in your answer. Do not spend too much time on the outline.

5. In the writing, always directly answer the question with the *first* sentence.

Example: Explain Pavlov's theory of conditioning.

Answer: Pavlov's theory of conditioning is based on . . .

The remainder is devoted to support by giving dates, examples, stating relationships, causes, effects, and research.

6. Present material that reflects the *grader's* personal or professional biases. Also, stick to the material covered in the reading or lecture and answer the question within the given frame of reference.

7. If you do not understand what the instructor is looking for, write down how you interpreted the question and then answer it.

8. If time does not permit a complete answer, use an outline form.

9. Write something for every question. When you "go blank" starting writing all the ideas you remember from your studying. One of them is bound to be close!

10. In sentence completion items, remember never to leave a space blank. When in doubt — GUESS. Make use of grammar to help decide the correct answer. Make the completed statement logically consistent.

11. If you have some time remaining, read over your answers. Frequently you can add some additional ideas which may come to mind. You can at least correct misspelled words or insert words to complete an idea.

12. Sometimes, before you even read the questions, you might write some facts and formulas you have memorized on the back of the test.

REFERENCE H

STUDYING FOR EXAMS

What to know before you start to study:

1. What type of test is it?

    a. Objective — multiple choice, true-false, matching, or a combination.

    b. Essay — short or long answer, or sentence completion.

    c. Problem solving.

    d. Combination of the above.

2. What material is to be covered?

3. How many questions (approximately)?

4. What is the time limit?

If the information above is not given by the instructor when he/she announces the test, ASK. This information is valuable to the way you study. Also, ask the instructor for old exams for your review.

STUDYING:

1. Be sure you have read all the material to be covered and have all the lecture notes before you begin your serious studying.

2. Plan what you will study and when.

3. Each review session should be limited to one hour. Take short breaks of 5-10 minutes between hourly sessions.

4. Try to predict exam questions. If it will be essay, try writing out the answers to your predicted questions.

5. Study in a group only if everyone has read the material. You don't gain as much when you must "tutor" someone else or if the other students are not prepared.

6. Prepare summary sheets from which to study to eliminate re-reading the textbook.

7. Review for objective tests by concentrating on detail and memorizing facts such as names, dates, formulas, and definitions (know a little bit about a lot).

8. Review for essay tests by concentrating on concepts, principles, theories, and relationships (knowing a lot about a little).

9. For problem solving tests, work examples of each type of

problem. Work them from memory until you get stuck. Then study your example problems and begin working them again from memory from the beginning. Do this until you can work the entire problem without referring to your notes.

10. The day of the test, do not learn any new material. This can interfere with what you have already learned.

11. Try not to discuss the test with other students while you are waiting to begin. If you have studied you do not need to be flustered by others making confusing remarks.

12. Try to consciously make yourself relax before the test begins.

13. After the test is over, do not waste time or energy worrying about how well you did. Start concentrating on your next exam.

14. Keep in good physical condition by maintaining a balanced diet and getting enough sleep.

REFERENCE I

MATHEMATICS LEARNING AND TEST-TAKING ACCOMMODATIONS FOR LEARNING DISABLED STUDENTS

Community colleges and universities are experiencing a substantial increase in learning disabled students. Improved special education programs in elementary schools, middle schools, and high schools have helped learning disabled students graduate. In the past, learning disabled students had special education classes with direct teacher instruction. Now, without separate special education classes, many learning disabled students are having major difficulties adapting to college level courses. In addition, mature students are making up a higher percentage of college students. Many of these mature students had previous learning problems in school, but existing programs could not diagnose them as learning disabled. Learning disabled students need additional learning skills and accommodations to reach their education potential.

Definition

The term, learning disability (LD), is used to describe a broad range of neurological dysfunctions. An LD is often misunderstood since it is invisible. The Federal Register defines an LD as:

A disorder in one or more of the basic psychological processes involved in understanding and use of language, spoken or written, which may manifest itself in an imperfect ability to listen, think, speak, read, write, spell, or to do mathematical calculations. The term includes such conditions as perceptual handicaps, brain injury, minimal brain dysfunction, dyslexia, and developmental aphasia. The term does not include individuals who have learning problems which are primarily the results of visual, learning or motor handicaps, or mental retardation, or environmental, cultural, or economic disadvantage.

A learning disability, according to the Title V regulations in the California Community College System is:

> ... a persistent condition of presumed neurological dysfunction which may also exist with other disabling conditions. This dysfunction continues despite instruction in standard classroom situations. Learning disabled adults, a heterogeneous group, have these common attributes:
>   a) average to above-average intellectual ability;
>   b) severe processing deficits;
>   c) severe aptitude-achievement discrepancy(ies);
>   d) measured achievement in an instructional or employment setting; and
>   e) measured appropriate adaptive behavior in an instructional or employment setting.

A specific learning disability, according to rule 6H-1.041 for the Florida Community College system, is:

> A disorder in one or more of the basic psychological or neurological processes involved in understanding or in using spoken or written language. Disorders may be manifested in listening, thinking, reading, writing, spelling, or performing arithmetic calculations. Examples include dyslexia, dysgraphia, dyscalculia, and other specific learning disabilities in the basic psychological or neurological process. Such disorders do not include learning problems which are primarily to visual, hearing, or motor handicaps, to mental retardation, to emotional disturbance, or environmental deprivation.

A general description of someone with a learning disability would be a person with average to above-average intelligence who has difficulties in one or more of the basic neurological functions such as perception, either auditory, visual or spatial. This disorder may impair reading (dyslexia), writing (dysgraphia), mathematical calculations (dyscalculia), thinking, and spelling. This disorder doesn't include learning problems due to physical disabilities, emotional disturbance, or lack of previous opportunity for learning. In some states the definition for a math learning disability would include a severe difference between mathematics or general aptitude scores and mathematics achievement scores. Learning disabilities can't be "cured." But in many cases learning disabilities can be circumvented through learning and testing accommodations.

Students, who may have a math LD, may have these symptoms:

Difficulty doing the actual calculations;
Difficulty learning a series of math steps to solve a problem;
Inability to apply math concepts to word problems;
Difficulty solving oral problems.

However, these same symptoms may be the results of other types of learning disabilities such as visual processing or auditory processing disorders.

Reasons for Math Learning Problems

Learning disabled students who have math learning difficulties may have a math learning disability. However, in my experience, most of these students have major information processing disorders. These processing disorders block their ability to obtain valuable information to learn math and/or to demonstrate their math knowledge on tests. Most of these learning disabled students did not have a severe difference between their general or mathematics aptitude and mathematics achievement scores. When the effects of their learning disability were diminished by appropriate learning and testing accommodations, their math grades improved.

As different processing disorders are discussed, pay special attention to the processing disorder that is evident in your learning disability. If you do not know your type of processing disorder, contact the counselor for disabled students for this important information. If you are unable to find out the type of processing disorder you have, then read the descriptions of the ones below. Pick the processing disorder that is similar to your learning problem. Remember: Only by going through an assessment program can you really know the type and extent of your processing disorder(s).

Learning disabled students who have visual processing speed disorders and visual processing disorders will have difficulty learning math. Such problems may include the visual processing speed that is the speed of working with understood math symbols and numbers. In other words, this is the speed at which you can copy down recognizable numbers and symbols. Your visual processing speed will affect how fast you can copy notes from the board and the speed at which you can take a math test.

Visual processing is your ability to recognize and remember, in sequence, complex math symbols and numbers that you may not know. For example, your ability to recognize and remember this polynomial: $4x^2 + 2x + 1$.

You may have difficulty telling the difference between 2 as a factor or an exponent, and between + as a plus sign, or x as a variable. Remembering the correct parts of the polynomial in order can be a problem. For instance, remembering 2x as the second part of the polynomial may become confused with 4x. This will cause major problems when you, the student, copy material from the board. It also will cause problems when reading tests and your math textbook. Mistakes can occur when you miscopy notes and misread the textbook or test questions.

Learning disabled students with auditory difficulties, including short-term memory and auditory processing, will eventually have difficulty learning math. Short-term memory difficulties are problems in remembering numbers, symbols or words in their correct order. This problem becomes apparent when listening to your math instructor explain the steps of working a problem.

For example: You may forget the math problem steps before you can write them in your notes. Also, you could recall and write down the math problem steps in the wrong order. Either mistake will cause difficulty in understanding the math problem and using your notes as a homework guide.

Students with auditory processing difficulties have problems telling the difference between certain sounds of words. These students will "miss" some of the words in a lecture or replace words with incorrect words. When this occurs, you will have difficulty understanding your instructor and writing your notes. Either auditory processing problem can cause you to misunderstand math concepts.

Learning disabled students taking math classes usually experience more difficulty learning math than their other courses. One major reason is that math requires linear learning. This means that material learned one day is used the next day and the next month. In many of your other nonlinear courses, material learned one day can be forgotten after the test.

Another reason for math learning problems may be your processing difficulties. These hinder you in receiving lecture information. The lecture information could be learned in the incorrect order or totally forgotten before the information is put into your notes.

For example: In some of your nonlinear courses — such as the social sciences — misunderstanding the first part of a lecture doesn't mean you won't understand the rest of the class material. However, in a math class, if you misunderstand the first concept taught, you are probably lost for the remainder of the class. In a nonlinear class, you can probably read the textbook to obtain the missed information. Due to the complexity of the math textbook, you probably won't be able to obtain that missed information from the textbook. These are several reasons you may have problems learning math compared to learning non-linear subjects.

Another type of learning disability is in reasoning or thinking. These problems usually occur when abstract reasoning is required to apply some type of mathematics concept from its law or principle to its application.

For example: You may have difficulty applying a formula to a new homework problem. You may remember the concept but forget how to use it in a problem. For the most part, you will have difficulty generalizing from an abstract math concept to your homework or test problems.

Some learning disabled students have difficulty with long-term memory. You may be inconsistent when learning new facts or concepts. You might be able to learn how to work fractions one day and a week later have difficulty recalling how to work fractions. However, when you are taught fractions again, it is easier for you to learn the steps.

The same memory loss can happen with your multiplication tables. You may forget the multiplication table, but you can demonstrate the concept of multiplication. You may have poor achievement in math calculations, but you may have average or above-average math reasoning ability. Usually, the result of a poor long-term memory is understanding how to work a problem but not having the mechanical calculation skills.

This discussion on reasons that learning disabled students have difficulties learning math has focused on specific learning disability disorders. Some learning disabled students have a combination of two or more learning disability disorders. Learning disabled students can also have additional learning problems that are a result of their learning disability. In general, learning disabled students may have problems in the following areas:

1) Understanding sarcasm;
2) Making polite conversation;
3) Developing appropriate study skills;
4) Planning time schedules (study, work);
5) Developing positive study attitudes;
6) Developing a positive self-image;
7) Concentrating with on-going distractions;
8) Retrieving words;
9) Following directions;
10) Having high test anxiety.

These learning problems are not characteristic of all learning disabled students. For example: I worked with a learning disabled woman whose major learning problem was visual speed. She also had a short-term memory problem. She had difficulty following directions and had high test anxiety. Her most positive learning aspect, like many other learning disabled students, was persistence and an achievement orientation for success. A student's learning disability does not just affect his/her education. It manifests itself in many signs and symptoms across the student's life.

Learning Accommodations

Learning accommodations allow learning disabled students the same access to course material as nondisabled students. Course materials can be in the form of lectures, labs, field trips, library resources, and books. Learning accommodations may be the same or different for various types of learning disabled students.

In my opinion, most learning disabled students with math learning problems have visual processing speed and/or visual processing disorders. In fact, with my learning disabled students, visual processing speed disorders are the most common math learning problem. Dr. Hessler, who wrote the book *Use and Interpretation of the Woodcock-Johnson Psycho-Educational Battery*, also suggested that visual processing speed is more highly related to mathematics achievement than other measures of perception. Many of these students are in the one-to-ten percentile in visual processing speed or visual processing. This means that 90-to-99 percent of the other students visually process material faster or better than the learning disabled student. For these students, taking math notes is a major problem. They usually take very few notes or try to write everything down. Learning disabled students who write down very few notes don't concentrate as much on the lecture and have little information for solving math problems. Other learning disabled students spend all their concentration on note-taking instead of listening to the instructor for a full explanation of the math concept. Both of these note-taking strategies leave the student with math-learning gaps.

Three lecture accommodations for learning disabled students with visual processing problems are note takers, tape recorders, and handouts. Learning disabled students may need only one or all of these learning accommodations. Learning disabled students may only need these accommodations in a math class and not in their other subjects. Accommodations depend on the learning disabled student's special needs.

When choosing a note-taker, try to get someone that is in your class. Have the potential note-taker give you a sample of his notes and ask him if he could adapt his note-taking style to meet your needs. The note-taker can use NCR paper, carbon paper or make a copy of his notes for your use. You still must take some notes to maintain concentration and use as a future guideline. Leave four or five lines between information gaps in your notes. Fill these gaps by recopying the notes from your note-taker. Try to copy the important information from your note-taker's notes to your notes on the same day. This procedure will help you understand and memorize your notes more quickly. The most important factor is to have good listening and note-taking techniques like those discussed in *Winning At Math*. You may want to teach your note-taker listening and note-taking skills as well!

You might want to use a tape recorder to preserve the lecture. Your tape recorder must have a tape counter to be useful. While taping the lecture, the instructor will start working a problem on the board. If this problem becomes difficult, write down the tape counter number and concentrate on the instructor's explanation. Once the instructor finishes the problem, then write down the tape number. As soon as you can, play back that part of the tape and complete your notes. If you still don't understand the problem, go to your instructor or tutor for help.

Remember: As a learning disabled student, you must rework your notes as soon as possible to get the most benefit from them. Use the reworking-your-notes techniques discussed in *Winning At Math*.

Handouts can improve math learning. Handouts can tell you the amount of note-taking necessary, list important information, and improve your concentration during lectures. Ask your instructor for future handouts or for additional handouts on current class materials. You can also make your own handouts by copying and enlarging math pages that have rules and concepts. Use these copies as a guide in the next math class.

Learning disabled students with visual processing disorders will have difficulty reading their math textbooks. Like most students, they use the same reading procedures for their math textbooks as they do for other textbooks instead of realizing that reading a math textbook needs a different approach. Reading a math textbook requires rapid recognition and an understanding of words intermixed with symbols. Students with visual processing problems become easily frustrated with math textbooks due to the various symbols and abrupt structure changes.

Another problem in reading a math textbook may be the failure to review previous materials before beginning new math textbook material. Textbook pages are so cluttered it is like going through a visual mine field, and many students don't survive. Many learning disabled students must be taught how to read a math textbook or have someone explain the textbook to them. However, math instructors lack the skills required to teach students reading techniques. Learning disabled students must develop reading techniques similar to the ones discussed in this book to understand their math textbook. To understand math concepts, learning disabled students must also learn the vocabulary and symbols that are specific to math.

A good reading technique can improve your understanding of the math textbook. Use the eight-step math reading technique discussed in Chapter Six to improve your understanding and retention of the math material. In some cases, an enlarged photo copy of a page with formulas may be helpful. Using different colored pens or magic markers to highlight different parts of equations that have different meaning can also be helpful. If these or other reading techniques do not work, then record the textbook on audio tape. However, a math textbook on tape requires a lot of listening time and can sometimes be ambiguous due to formulas and graphs. To improve your learning of the math textbook, listen to the audio tape as you read the textbook.

In many cases, a tutor will be required for students with visual processing problems. Even with the already mentioned learning accommodations, a tutor will have to go over some of the more difficult math concepts or problems such as graphing equations.

The tutor should use different colored pens or highlighters to indicate exponents, x and y axes or sign changes. The tutor should also have some training on the characteristics of learning disabled students. The tutor should write with a felt-tipped pen or do problems on the chalkboard to make it easier for the learning disabled student to distinguish between factors, numbers and variables. If possible, have tutors ask partially sighted students some of their learning strategies which may help learning disabled students with visual processing problems.

Learning disabled students with auditory disorders, including short-term memory and auditory processing, are the second most common group of learning disabled students with difficulty in math classes. If you have a short-term memory problem, you may not remember the math steps in order, or you may forget certain problem steps. You may remember most of the words but get them mixed up with other words or have gaps in your memory. You may not be able to remember facts, understand concepts and write them down all at the same time. Your notes may have the problem steps in the wrong order, or you may not remember the problem steps long enough to put them in your notes. Since math is considered linear learning, misunderstanding one word or not remembering one word of a concept or formula can cause a misunderstanding of the total concept or formula. The result is that you are only receiving part of the lecture. This leads to gaps in your notes and learning.

Some of the learning accommodations for learning disabled students with auditory processing difficulties are note-takers, tape recorders, physical proximity and video tapes. As discussed earlier, the note-takers should be from the same math class or at least have the same math teacher. The learning disabled students should be able to understand the note-taker's notes and use these notes to fill in the gaps in their own notes. The learning disabled students should still take some of their own notes; they should not rely only on the note-taker's notes. Learning disabled students must rework their notes as soon as possible after class to improve their learning. While reworking the notes, write down questions about unclear material to ask the instructor.

The major problem in using an audio tape recorder is the listening time. Listening to an hour math tape might take an hour and a half. Using an audio tape recorder with a tape counter can help learning disabled students improve their understanding of lecture material with less wasted time. The tape counter indicates how much of the audio tape has passed through the tape player. Students can use the tape counter to indicate the tape number of the beginning and end of a math problem discussion. Later on, that exact discussion can be located on the tape and replayed. This will allow students to fill in the gaps in their notes and to better understand the material. Listening to the entire tape may frustrate learning disabled students with auditory processing difficulties. This is the reason for using tape counter numbers to locate the difficult parts of the lecture.

The use of video math tapes as part of the textbook publisher's package, video tapes for the learning disabled, or made-by-the-math-department tapes can be helpful. These video tapes can be especially helpful if the tapes have subtitle captions. Students with audio problems usually have a visual preferred learning style that is conducive to learning from video tapes.

Video tapes can be played backwards and forwards as many times as needed to understand the material. Students can take notes from the video tape at their own speed. An excellent learning strategy is to watch the math video tape on a difficult math concept right after math class.

Trained tutors will be required for some of the learning disabled students with auditory processing/short-term memory problems. Tutors must understand that these learning disabled students will probably only understand part of the tutor's problem-solving explanation at any given time. Tutors should have the learning disabled students repeat back the just-explained facts or concepts. This is one way to make sure learning disabled students understand the facts or concepts.

Learning disabled students with thinking/reasoning problems will have extreme difficulty learning math. This could also include students with head injuries. These students usually have poor organizational skills, poor problem-solving skills, and trouble understanding causal relationships. The difficulties causing the most problems are poor abstract reasoning and difficulty generalizing from one experience/idea to new situations. If this is your learning disability area, you may learn how to do a math problem but not be able to generalize that concept to your homework problems. You may demonstrate the knowledge of a math concept one day and forget how to use the same math concept the next day. This usually occurs when you try to memorize math problem patterns because you cannot understand the math concept. It may be difficult for you to learn mathematics, but I feel you should go as far as you can in your math courses.

Learning accommodations for learning disabled students with thinking/reasoning problems are extensive. Most of these students will need a combination of note-takers, tape recorders, handouts, math video tapes, tutors and calculators. Tutors who are trained in helping learning disabled students are the most important learning accommodation. Tutors must start with learning disabled students no later than the second week of class. The tutor should start with a review of the material, then the tutor can start on the new material. Before ending the tutoring session, a review of the new material must be conducted. By following this tutoring procedure, the learning disabled student will retain more but learn less per tutorial session.

Remember: Working with students with thinking/reasoning problems will require more patience and tutorial sessions because less is learned in each session.

Calculators can be another learning accommodation for learning disabled students with thinking and reasoning problems. Calculators can reduce some of the problems with the mechanics of math. This will allow the student to have a better focus on math concepts. Many learning disabled students working math problems don't know if they got the problem wrong due to their calculations or because they have copied an incorrect math concept. Calculators can help alleviate this problem so that students can focus on applying the correct math concept instead of worrying about the mechanics.

Some students have difficulty with math due to a long-term memory retrieval problem. Students may appear to know their basic math skills one day and the next day forget how to do a basic calculation. If this is your learning disability area, then you may score good grades on quizzes but fail major tests. When you are taught a forgotten concept again, you relearn it faster than the first time. This means that you retained some of the concept knowledge but not enough to remember the total concept. This long-term memory problem is different than thinking/ reasoning problems. A long-term memory problem mainly applies to the mechanics of doing math problems or not remembering a concept. Being told the math concept should allow you to work that type of problem.

If you have this type of learning disability, then some learning accommodations can be math video tapes, math study skills, note cards, calculators, and tutors. You will have to constantly review difficult math procedures to keep from forgetting those procedures. This can be accomplished by marking certain parts of math video tapes or using note cards for review. Calculators should be used to take the place of the mechanical aspects of math. Tutors can be used to help you review the parts of math that you keep forgetting. Before every math test, you must understand the concept errors of the previous test and know how to use these concepts.

This discussion has focused on learning accommodations for specific types of learning disabilities. You may have problems in one or more of the processing areas. This means a combination of learning accommodations may be needed to enhance your learning. Like other learning disabled students, you may have additional learning problems, including time management, high test anxiety, social interaction difficulties, and poor test-taking skills. You need to focus on these additional learning problems as much as your learning accommodations.

In summary, having processing deficits or having dyscalculia is only one of many learning disabilities that can cause students to have problems learning math. Most learning disabled students who have difficulty in math have visual or auditory processing disorders. These processing disorders can cause an interruption or complete disruption in obtaining lecture material inherent to learning math. Learning disabled students with thinking/reasoning problems will have the most difficulty learning math because of their poor ability to grasp math concepts. Learning disabled students with long-term memory problems will have difficulty remembering their basic math skills or the math concepts. However, in many cases, learning disabled students with the proper accommodation and good math study skills can succeed in math.

Testing Accommodations

Appropriate testing accommodations for learning disabled students are used to separate measuring a student's learning disability and measuring the student's math knowledge. This is especially true when time is considered a major factor in measuring math knowledge. In the past, when some learning disabled students were given a math test, the test measured both the degree of their learning disability and their math knowledge. In almost all these cases, the result was a lower math grade, which was not a true indicator of the student's math knowledge. This grade frustrated the learning disabled math student and in many situations frustrated the math instructor. The student was frustrated because he/she knew more than the test indicated. The math instructor also became frustrated because the student demonstrated math knowledge in class but failed the test. This section will suggest appropriate testing accommodations for math students with different types of disabilities. This is not an all-inclusive list but can be considered an appropriate guide.

Most of the learning disabled students I have helped who had difficulty in math had a visual processing and/or visual speed problem. If this is your major type of learning disability, then timed math tests will cause major problems in determining your math knowledge. Let's say you start taking a math test with fifteen problems. Most math students will be able to read each math problem in a matter of seconds. But it will take you a lot longer. In fact, it might take up to a minute to decipher a complicated math problem, and you still may have misread the signs. While the other students are working on the first math problem, you are still making sure you read it correctly. The extra time for reading math problems may cost you an extra ten minutes.

After reading the math problem, you start working on the answers step-by-step. However, it takes you longer to work the problem due to slower visual speed and/or your need to constantly recheck your work for misread symbols or signs. You are now on problem eight and have only eight minutes left to complete the test. You decide to rush through the remainder of the problems. You make careless errors and do not finish the last four questions. Unfortunately, you knew how to work two of the last four problems. Taking more time to work the math problem and checking each step may cost you an extra fifteen to twenty minutes. Your test is returned with a grade of 58. You failed the test. But remember, half of your test time was taken up coping with your learning disability. Was this test a true measurement of your math knowledge? NO!

You need to have as much extra time as needed to compensate for your learning disability. Depending on the type and severity of your learning disability along with the type of test, you may need one and a half to two times the normal time. However, I allow some learning disabled students up to three times the normal time, if they have a severe visual speed/processing problems and are taking math tests with graphing problems.

Another testing accommodation for learning disabled students with visual speed/processing problems is to enlarge the test questions. By enlarging the test and spreading out the questions, you are less likely to become confused by similar symbols and numbers. This can also be accomplished by having the math instructor use a felt-tipped pen to handwrite the test in large numbers and symbols.

A second method is to enlarge the test by blowing up the test 150 percent and putting two questions per 14" by 17" sheet. You now will have less of a chance of misreading the test question and plenty of room to work the problems. You should then divide the working space of the test page in half. On the left-hand side of the page you work each step of the problem. The calculations that are required for each problem step are put on the right-hand side of the page. Using this system will be less confusing because you will not mix up problem steps and calculations. Make sure you check every answer of each math problem before turning in your test.

Test readers and/or test proofreaders are two or more testing accommodations that can compensate a math student with visual speed/processing problems. You may chose a test reader to make sure you are correctly reading the test. When you receive the test, you need to read the test out loud to the test reader to compare what you read to each problem statement. If you make a mistake, then that part of the problem should be written in large print in a different color. If you do not understand the test question, then the test reader can clarify the meaning of the test questions. When clarifying the test question, the reader has to make sure not to include information on how to work the problem. As you finish each problem or after finishing the entire test, the test reader can become a test proofreader. The test proofreader will read each problem, step by step, as you have written it. If you disagree with what the proofreader said, then you need to review that problem.

When I was a test proofreader, students usually found at least two major mistakes, most often in the form of missing sign changes. Many of these students misread signs in the problem steps. If you have a history of reading math problem steps one way, then reading it a different way, the next time you will need a test proofreader. Just using a test reader may present a problem. You may read the math step a different way the second time leading to a wrong answer!

Recording the test questions on an audio cassette tape and playing the tape while you read the questions is another testing accommodation. By hearing the test question on tape, you will be able to make sure you understand the numbers, symbols and words of the math problem. If your main problem is just reading the test question and not misinterpreting your own handwriting, this could be your best testing accommodation.

You may require a private testing area with or without a

chalkboard. If you have some problems with distractions while testing, you may need a private testing area. If you are working with a test reader or test proofreader, then you will need a private testing area so as not to disturb other students.

Some students can think better when walking around and writing problems on the chalkboard. By writing problems on the chalkboard, there is less chance of misreading signs and working the problem incorrectly. Moreover, you most likely learned how to work the problem from a chalkboard. If this is true, then recalling how to work the problem may be easier standing at the chalkboard.

You may have difficulty keeping the steps of a math problem in order, unless you have lined scratch paper. You can use regular lined paper or get it blown up to 125 percent to make it easier to write on. Another way to use lined paper is to turn it on its side to make the lines into columns. Put a number, sign or variable of the equation in separate columns. Use the columns to keep the numbers, signs and variables in order. As you work the problem, the columns of numbers, signs and variables will decrease. In most cases, you should end up with two columns being used that contain a variable and number.

These are a few testing accommodations for learning disabled students with visual speed/processing problems. This does not mean that you will need all these testing accommodations to circumvent your learning disability. The combination of testing accommodations, I usually recommend is extended test times, enlarged tests, and a private testing area. These three testing accommodations have solved most of my learning disabled students' testing problems.

Test accommodations for learning disabled students with auditory processing and/or short-term memory problems depend on the type and length of the math test. If the math test contains some type of oral component, then you will need a private testing area with a test reader or tape recorder. If you have a short-term memory problem, then math tests with story problems, having questions with several parts or having long written introductions will take more time to complete. You may have to read the test questions many times or return to different parts of the questions to obtain a full understanding. Meanwhile, the other students are already working on the test problems.

These other students will most likely finish before you do and have time to check their answers. You may have taken an additional ten minutes to reread and remember the questions, leaving less time to finish to the problems and check your answers.

Learning disabled students with thinking/reasoning problems or long-term memory problems will need special testing accommodations. If this is your type of learning disability, then test-taking time will be a major problem. It will probably take you longer to remember how to work some of the problems or to do the mechanics of working the problems. You may also need a calculator to work the problems in order to make sure that your calculations are correct. This will increase the chances of using the correct concept because of less chance of careless errors. Using the Strategy Cards for Higher Grades suggestions and putting down your mnemonic and acronym memory cues are also important.

There are different types of testing accommodations for different types learning disabilities. Some learning disabled students may not require any testing accommodations where other learning disabled students will require extensive testing accommodations. Make sure you discuss which appropriate test-taking accommodation you will need with the disabled student staff at your institution. Practice explaining the reasons for the test-taking accommodations with your instructor. This practice will prepare you in case your instructor asks you why test-taking accommodations are necessary. Let your instructor know several weeks ahead of time about the testing accommodations you will need.

Sometimes you or your counselor may not know the effects of your learning disability and what testing accommodations you should receive. In this situation, I would suggest the following testing accommodation: private testing area, extended time, and enlarged test. You then must find out your type of learning disability to make sure that you are receiving appropriate testing accommodations.

Alternate Test Forms

In most cases, the instructor will select the test form that he/she thinks will be the most effective way to evaluate a student's mathematics skills. The test formats are usually based on tradition and easy to score. Unfortunately, some of the test forms will cause difficulty with certain learning disabled students. Even though large-print tests and audio cassette tape test forms have been discussed, there are other different test forms that could be used. Some of these other test forms are: oral tests, video monitor enlarged tests, computer enlarged tests, computer speech enhanced tests, and take-home tests. Each test form has its advantages and disadvantages for the student and instructor. You need to discuss these alternate test forms with your disabled support staff to find out which alternate test form is likely to give the best indication of your math knowledge rather than your learning disability.

Case Studies

This next section is about learning disabled students whose math grades have been improved. In my opinion, most learning disabled students need a combination of math study skills, learning accommodations, testing accommodations, and proper test format in order to reach their math learning potential. Some learning disabled students only needed the Math Study Skills course, using the *Winning At Math* textbook and the audio cassette tape, *How To Reduce Test Anxiety*, to become successful in math. Remember: Math study skills are especially important because you are already having difficulty receiving/processing lecture/textbook information. Math study skills learning techniques can compensate for some of your learning problems. It will not do you much good to have all the accommodations requested if you do not have good math learning and test-taking skills. On the other hand, without the appropriate testing accommodation and test form, you cannot demonstrate your math knowledge.

The following case studies are an example of combining math study skills, learning accommodations, testing accommodations, and appropriate testing forms to help students become successful in math. I will proceed from the least difficult to the most difficult case histories. These are only a few case histories of learning disabled students with whom I have worked to improve their math skills.

Student A is a thirty-year-old male who was in my Math Study Skills course and was having difficulty passing his basic mathematics tests. It appeared from his class participation and homework assignments that he was using the correct math study skills and test-taking skills. While talking to him, he indicated that he had a D average on his math tests and did not have time to finish his math tests. He said that it took a lot of time to write down each problem step, and he constantly checked for careless mistakes. Doing his math homework took him longer than most students, but it appeared that he understood the math concepts. He also said that in high school he had reading problems and so did some of this other family members. Reviewing his math test revealed that the completed problems were mostly correct. The problems toward the end of the test had careless errors or were not completed.

Student A's intelligence test score indicated that he had average intelligence. *The Woodcock-Johnson-R Test of Cognitive Ability* indicated his visual processing speed test score was extremely low. The visual processing speed score was at the fifth grade level, which is the second percentile. This means that 98 percent of the other students had better visual processing speed skills than Student A. A fifty percentile score is a middle score, meaning that half of the other scores are below and half are above that score. There was over a thirty-point difference between his visual processing speed and his IQ. His visual processing (which is separate from visual processing speed) grade equivalent was 16, which is at the 73 percentile.

The test results indicate Student A had difficulty working quickly under pressure without making careless mistakes. This meant that he would take more time on math tests and made more careless errors. On the other hand, Student A had good skills in recognizing figures, finding incomplete figures and understanding spatial configurations. He could quickly recognize complicated math equations and formulas.

As you may remember, a learning disability in Florida is: "A disorder in one or more of the basic psychological or neurological processes involved in understanding or in using the spoken or written language. Disorders may be manifested in listening, thinking, reading, writing, spelling, or performing arithmetic calculations."

Student A had a learning disability because his visual processing speed disorder dramatically decreased his writing speed and accuracy of written material. This problem manifested itself in not performing fast arithmetic calculations during tests and constant rechecking of problem steps. It also blocked his ability to understand math problems written on the board because his concentration was on accurate note-taking instead of the instructor's explanations on working problem steps.

The accommodations Student A received were note-takers, private testing area, and extended test time. Student A made a B on his next math test and an A on his final math test for a B math course grade. He is now in the next level math course and making good grades. His case is a good example on how the combination of math study skills, learning accommodation, and testing accommodation resulted in a student becoming successful in math instead of failing the course.

Student B is a middle-aged woman who had problems with concentration, distractions, test anxiety, reversing letter/ numbers, and short-term auditory/visual memory. She indicated a lifelong history of math learning problems. She also said that many of her family members had similar learning problems. With all these problems, she had a high grade point average. She was taking her first math class in about twenty years. She was in the basic math class for a while and decided to drop the course due to anxiety and learning problems. She was referred to a learning center to improve her basic math skills and continued to take the Math Study Skills course.

An intelligence test score indicated she had an average intelligence. The scores from the *Woodcock-Johnson-R Test of Cognitive Ability* indicated a severe visual processing speed learning problem and a possible auditory processing learning problem. Even though she had average intelligence, she had a fourth grade level visual processing speed score or a 0.1 percentile score. This means that over 99 percent of the students had a better visual processing speed than she did. Her auditory processing was at the fifth grade level or a 17 percentile score. This means that 83 percent of similar students had better auditory processing than she did. However, her visual processing was at the grade 13 level or a 62 percentile score. There was not a significant difference between her math achievement score and her intelligence score.

Student B was learning disabled because her visual processing speed disorder prevented her from accurately writing material as fast as other students. This problem manifested itself in not performing arithmetic calculations as quickly as other students during tests. This caused her problems on timed math tests. It also blocked her from understanding math problems written on the board. She took more time to copy the material off the board, leaving less time to understand the math problem. Her low auditory processing score suggested a low ability for understanding sound patterns (words) under distracting conditions. She also had difficulty in putting sounds together to make words. Student B had problems understanding lectures in a noisy classroom and confused one word for another word. She was having difficulty taking notes. She left out words in her notes or put the wrong words in her notes, causing problems in understanding math concepts.

The accommodations for Student B were a note-taker for her math course, extended test time, and a private quiet testing area. Student B made an A on her first math test, which was the first A she had ever made on a math test. A combination of previous math learning, math study skills, appropriate accommodations and persistence had made her a successful math student.

Student C was dyslexic and extended test time was suggested as an accommodation. Student C performed poorly on her first major math test. After having an interview with Student C, additional testing was conducted to reveal the aspects of her learning disability. Since she was not in the Math Study Skills course, additional affective surveys were given. Student C had test anxiety at the 99 percentile, excellent general study skills and study attitude, external locus of control, and an auditory numeric learning style. To decrease the test anxiety, she was instructed to practice with the audio cassette tape, *How To Reduce Test Anxiety*. She was also given instructions on the best procedures for auditory learning.

Student C had average intelligence but an extreme disorder in visual processing speed, short-term memory, and visual processing. Her processing speed grade level was at the fifth grade level, which is at the first percentile. Her short-term memory was at the second grade level, which is at the third percentile. Her visual processing was at the third grade level, which is at the fourth percentile. Student C had a learning disability in visual processing speed, short-term memory, and visual processing.

The effects of visual processing speed on learning and test taking have already been discussed. Student C's short-term memory deficit caused problems in storing information and using that information. For example: If a math instructor explained five steps to solving a math equation, she had difficulty remembering the steps long enough to write them down. Her visual processing ability was also a concern because of her difficulty in recognizing the meaning of a combination of math symbols and numbers. Student C had major problems remembering, understanding, and writing down math concepts in the form of notes.

The accommodations for Student C were expanded to include a note-taker and a tutor. The note-taker allowed her to concentrate more on the math instructor's explanation while still taking some notes. The math tutor explained the math concepts on an individual basis to enhance Student C's learning. With these accommodations, Student C made a B in her math course and is making good grades in her next math course.

The next student case presented the most challenge in helping him to learn math. At the suggestion of Bill Reineke, math lab director, I am calling this student Anti-X instead of Student D. Bill said that Anti-X was an appropriate name for this student because of his dislike for math. "Anti" represents what a person dislikes, and the symbol "X" represents math.

Three years ago, Anti-X was referred because of his math learning problems. He was failing one of the basic math courses. His major was computer science, and he was making A's and B's in all his computer programming courses. I remember him indicating several times that he did not like math. He wanted to know why he needed math since he was making good grades in his computer courses. After talking to him about math learning skills, I referred him to the next Math Study Skills course. Since learning math was similar to learning computer programming, I thought math study skills and test-taking skills were his main problem. With the help of the Math Study Skills course, he passed the math course.

The following semester he enrolled in the next level of math and began to have learning difficulties. Anti-X returned to my office saying that he was studying math about twenty hours a week but was failing the course. However, he was still making A's and B's in his computer courses. I decided that there must be something else affecting his ability to learn math.

Anti-X completed a questionnaire for a learning disability evaluation. During the interview, he indicated some motor nerve impairment which slowed down his handwriting. He also indicated problems with visual perception speed and memory problems. He indicated that the main problem was putting math concepts into long-term memory. As a side comment, he said that he became disoriented in new places until he learned his way around. His grade point average was a 3.22.

Anti-X completed the *Survey of Study Habits and Attitudes*, *Math Study Skills Evaluation*, old *Woodcock-Johnson Psycho-Educational Battery (WJPETB)*, and the *Wechsler Adult Intelligence Scale-R*. Anti-X had a score of 95 percentile and 99 percentile respectively on *Study Habits and Study Attitudes*, which indicated excellent study skill and attitudes. He made a score of 82 percent on the *Math Study Skills Evaluation*, which indicated good math study skills. There was a 23 point difference between his Verbal IQ and Performance IQ. His Verbal IQ was the highest and was used as the best indicator of his ability. His Verbal IQ was in the average range.

The old *WJPETB* combines the visual speed and visual perception test. Anti-X's visual speed and perception indicated a 4.0 grade level score, which represents the first percentile. Anti-X's reasoning score was low, with a 2.0 grade level score, which is third percentile. The reasoning score measures nonverbal abstract reasoning and problem-solving skills. The memory score of the old *WJPETB* measures short-term auditory memory. Anti-X's memory grade level score was 12.9, which is at the 64th percentile. There was no significant difference between Anti-X's mathematics aptitude score of 85 and his mathematics achievement score of 92. Based on this data, Anti-X should not have passed his first math course or be making A's and B's in his computer programming courses.

Due to his documented motor nerve impairment, Anti-X was classified as having a physical disability. However, the testing data indicated that he had a learning problem due to the severe difference between his IQ score and his visual speed/perception score. If Anti-X had not been classified as being physically disabled, he would have qualified under Florida law as a learning disabled student.

Learning accommodations for Anti-X challenged the best of both of our abilities and imaginations. Learning accommodations changed several times based on different math concepts and math courses. Anti-X did not use all of these learning accommodations at the same time but selected the learning accommodation that was best suited for learning the math concept.

Anti-X was having difficulties keeping his math homework in some legible organization. He would bring his homework problems for the tutors to review, but the tutors couldn't understand what he wrote. Anti-X improved his homework technique by drawing a line down the center of the page. On the left side of the page, he wrote the problem steps, and on the right side of the page, he did the mathematical calculations for each step. This kept the problem steps separate from the calculations.

Anti-X misread some of the mathematical notations and signs while doing his homework. I had Anti-X use a felt-tipped pen which made the math numbers and symbols larger and easier to read. Anti-X used this system for a while but abandoned it for another system. Anti-X now used a ballpoint pen that had four colors, red, black, green, and blue. When doing math homework, he used different colors for different variables, numbers, factors, exponents, negative signs or positive signs. This does not mean that he has a different color for each part of the math problem. He used blue for part of the math problem he already knew how to work and used the other pen colors for the new parts of the math problem. Now he could follow each math operation by pen color until the problem was completed. Even though it took Anti-X longer to work the problems this way, it decreased his visual processing errors and made it easier to work the problems.

Anti-X worked with several math tutors as an additional learning aid. The best tutor Anti-X had was a very intelligent visually impaired tutor named Mike Farnham. Mike had already developed different techniques for learning and taking math tests that circumvented his visual disability. Mike shared these techniques with Anti-X. For example, we discovered that Anti-X's writing speed was not capable of keeping up with the speed he mentally processed math problems. When Anti-X's mental processing speed got ahead of his writing speed, he made careless errors and lost track of the problem-solving procedures. Anti-X was asked to solve this problem in two ways. One way to reduce his mental speed was to write each math problem step in a different colored pen. This reduced Anti-X's mental problem-solving speed, resulting in fewer problem steps omitted. This procedure was then combined with having Anti-X write, in English, what he was doing beside each math step. This procedure not only slowed down Anti-X's mental-processing speed but also helped him learn the reasons for each problem step.

Teaching Anti-X how to do math graphing problems was another challenge. Mike determined the best method for Anti-X to learn graphing was using the "shift method." Rather than merely plotting points and drawing a line through them, Mike gave Anti-X a concept model about parabolas. Anti-X was taught how to draw a simple parabola on a graph. Using the standard parabola equation of $y = (x - h^2)$, Anti-X was taught that the sign inside the parentheses determines if the parabola is to the left or right of the y axis. A negative sign shifts the parabola to the right and a positive sign shifts the equation to the left. To better understand the graphing of a parabola, Anti-X drew a basic parabola. He then changed pen colors and drew a new parabola with a shift to the right. He changed pen colors again and drew another parabola with a shift to the left. The different colored parabolas made it easier to understand graphing parabolas. By looking at the sign of h, Anti-X could tell which side of the y axis to put the graph on. Finally, these graphs were put on index cards with different pen colors to be understood and memorized.

Anti-X was also having difficulty solving problems involving complex numbers such as:

$$(3a^4b^2c^{-2})^{-4} \ (2a^{-2}b^3c^2)$$

He would look at the total problem and become confused and not know where to start. To solve this problem, a rectangular piece of cardboard was cut out to lay over the first part of the problem. This allowed Anti-X to isolate different factors or signs and concentrate on one part of the problem at a time. This eliminated most of the visual confusion caused by his learning disability.

Anti-X had difficulty understanding and working word problems. Anti-X was taught to break down the work problem into two parts. He wrote down the word phrases for each part of the word problem and put the algebraic expressions under each word phrase. For example: If the word phrase was "a number decreased by seven," then he wrote "x − 7" under that word phrase. Anti-X used the Strategy Card Number 12, "Translating

English Words Into Algebraic Expressions," and learned how to write the algebraic expression for word problems.

Many different learning accommodations were tried with Anti-X to make the mathematics material accessible to him. Not all these learning accommodations were used at the same time. Some other learning accommodations were tried but did not work. The team approach by Anti-X, his tutor and myself successfully developed learning accommodations that helped Anti-X improve his ability to learn mathematics.

Anti-X's testing accommodations were changed several times. Basically, this included extended test time, private testing area, and test proofreaders. Some of his testing accommodations were similar to his learning accommodations. Since Anti-X had difficulty with his handwriting due to a motor nerve disability and a visual speed/perception learning disability, he was given three times the normal testing time. The private testing area was in a comfortable room with a chalkboard. He was allowed to take breaks under supervised conditions. This allowed him more control over his anxiety. Sometimes test proofreaders were used to read the test to him and the answers back to him. Later on, test proofreaders were discontinued.

Using an alternate test form was a major breakthrough for Anti-X. At first, we used the regular teacher-made test with a reader, but the test format was so visually crowded that he had difficulty understanding the problems and writing the steps to the problems. On several occasions, Anti-X miscopied problems he had solved on scratch paper back to his test. I then had the test blown up 150 percent so that four problems were on a page. This helped Anti-X but did not totally solve the visual processing problem. Anti-X's test was then blown up 150 percent with one or two problems per 14" by 17" page. This test format eliminated the misread problem errors and eliminated the need for a test proofreader.

However, another problem occurred when Anti-X mixed up his steps in solving math equations with the calculations used to solve the equations. For example: After working three fourths of a math problem, Anti-X could not find the last step used to solve the equation. His calculations were so mixed up he could not find where he left off working the equation. To solve this problem, I had Anti-X divide his test page in half by folding it in the middle. The left side of the page was to work the problem step by step. The right side of the page was to do the calculations. This test-taking strategy solved the problem of mixing up the problem steps and calculations.

When Anti-X began to have tests on graphing equations, his visual process/speed disability became a major problem. Anti-X could not tell the difference between the points on the X and Y axes. To solve this problem, regular graph paper was blown up 150 percent with one sheet of graph paper per problem. On the graph paper, Anti-X first drew the X axis in red, then drew the Y axis in blue. Each point on the X axes were indicated by a red mark. Each point on the Y axes were indicated by a blue mark. This marking system made it possible to locate the coordinates on the graph. After having several tests with graphing problems, Anti-X only needed to draw in the X and Y axes with different colors to mark the coordinates.

Anti-X next took a statistics course. His statistics course was causing him some visual problems, especially with bar graphs. To solve that problem, Anti-X made each bar graph and the measurements of the bar graph a different color. He used graph paper that was enlarged 150 percent for his test. To better understand statistics formulas, Anti-X used different colors to distinguish different parts of statistics formulas. Anti-X now has a "B" average in his statistics course.

Currently, Anti-X has finished all three of his required algebra classes and is now taking the statistics course. Anti-X has become a successful math student through persistence, good math instruction, and accommodations. In fact, Anti-X has become so successful that Bill Reineke asked him to be a math tutor. Since Anti-X now likes math and has become a math tutor, I have dropped "Anti" from his fictitious name. He can now be called Student X.

Some people might question how Student X learned college algebra with his reasoning at a second grade level. In reviewing the four subtests that make up the reasoning cluster, two of those tests require a high level of visual processing. These two reasoning subtest scores were below the two other subtest scores. In other words, the reasoning test was also measuring his poor visual perception/speed instead of just measuring his reasoning ability. This lowered his reasoning score and was not a true indication of his reasoning ability. This one incident does not mean that all the reasoning scores are biased for students with poor visual processing.

A learning disability is a neurological disorder that can not be cured. Learning disabled students must be given accommodations to make sure they have appropriate access to learning materials. Learning disabled students also need testing accommodations to measure their achievement so it is not affected by their learning disability. This right is guaranteed by Section 504 of the Rehabilitation Act of 1973.

Classroom instruction also can be altered to help learning disabled students improve their learning. Research suggests that altering teaching styles to accommodate learning disabled students can also improve the learning of nondisabled students. However, in some cases learning disabled students, even with all the suggested accommodations, may not pass the required mathematics courses to graduate. These students, who have advanced as far as they can in mathematics, can be allowed course substitutions. Some of the course substitutions for mathematics are: business mathematics, computer sciences, general sciences, logic, and accounting. The computer science courses can be a course which teaches students how to run computer programs such as word processing programs. The logic course is usually taught by the Philosophy Department. The course selected for substitution should have the least effect on the student's learning disability and be closely related to the student's major or concepts reasoning mathematics. Consult the Disability Support Service Office for more information about the rules on course substitutions.

INDEX

224

FIGURES

BIBLIOGRAPHY

Bloom, B. (1976). *Human Characteristics and School Learning*, New York: McGraw Hill Book Company.

Fox, Lynn H. (1977). Women and Mathematics: Research Perspective for Change. Washington, DC: National Institute of Education.

Ieffingwell, B.J. (1980). Reduction of test anxiety in students enrolled in mathematics courses: Practical solutions for counselors. Atlanta, Georgia: A presentation at the Annual Convention of the American Personnel and Guidance Association. (ERIC Document Reproduction Service No. ED 195-001).

Richardson, F.C., and Suinn, R.M. (1973). A comparison of traditional systematic desensitization, accelerated mass desensitization, and mathematics anxiety. *Behavior Therapy, 4,* 212 - 218.

Tobias, S. (1978b). Who's afraid of math and why? *Atlantic Monthly*, September, 63-65.

Wolpe, J. (1958). *Psychotherapy by reciprocal inhibition*. Stanford: Stanford University Press.

AUTHOR BIOGRAPHICAL DATA

Over the past 15 years Learning Specialist Dr. Paul D. Nolting has helped thousands of students and professionals increase their math test scores and obtain better grades — easily and quickly.

He is the author of two innovative books, *Winning at Math* and *The Effects of Counseling and Study Skills Training on Mathematics Academic Achievement*, and two cassette tapes "How to Ace Tests" and "How to Reduce Test Anxiety."

Dr. Nolting holds a B.S. degree in Psychology, an M.S. degree in Counseling and Human Systems from Florida State University, and a Ph. D. degree in Counselor Education from the University of South Florida. He is also licensed as a Counselor and a National Certified Vocational Evaluator. He is an instructor at Manatee Community College and the University of South Florida, president of the Ability Discovery Center, Bradenton, Florida, and a member of the Board of Directors of the Florida Developmental Education Association.

Dr. Nolting has also consulted with colleges, helping them improve mathematics academic achievement for their students.

A key speaker at numerous regional and national educational conferences and conventions, Dr. Nolting has been widely acclaimed for his ability to communicate with students on the subject of improving grades.

Dr. Nolting has conducted national pre- and post-conference workshops for the National Association of Developmental Education and the Association on Handicapped Student Services Programs in Post-Secondary Education. He has conducted national training workshops for the National Council of Educational Opportunity Associations and the Association on Handicapped Student Services Programs in Post-Secondary Education. His most recent honor was the Florida Association of Community Colleges, 1990 Student Development Commission Exemplary Practice Award. This award was for developing a plan for reducing mathematics attrition called, "Math Learning Problems: Diagnosis and Strategic Learning Plans."

ACADEMIC SUCCESS PRESS

Light Years Ahead in Learning

Dear Educators, Students and Concerned Parents:

We at the Academic Success Press are dedicated to developing education materials to help students improve their ability to learn and obtain better grades.

The principle author and consultant to the Academic Success Press is Learning Specialist Paul D. Nolting, Ph.D. Over the past 15 years, Dr. Nolting has helped thousands of students and professionals improve their ability to learn and obtain better grades — easily and quickly. He is the author of two valuable cassette tapes . . .

HOW TO REDUCE TEST ANXIETY and *HOW TO ACE TESTS*
and the innovative book
WINNING AT MATH
Your Guide To Learning Mathematics
Through Successful Study Skills

Dr. Nolting holds a B.S. degree in Psychology and an M.S. degree in Counseling and Human Systems from Florida State University, and a Ph.D. degree in Counselor Education from the University of South Florida. He is also licensed as a Counselor and a National Certified Vocational Evaluator. He is an instructor at Manatee Community College and the University of South Florida, president of the Ability Discovery Center, Bradenton, Florida, and a member of the Board of Directors of the Florida Developmental Education Association.

Dr. Nolting has conducted classes and workshops for students at numerous high schools and colleges, helping them to improve their level of academic achievement and reduce their test anxiety. He has also trained high school and college faculty members on how to effectively teach study skills courses and how to introduce their own study skills programs in their various institutions.

For example, one of Dr. Nolting's very successful study skills programs is a one-hour accredited mathematics study skills course for students taking math preparatory courses. The course and the course cassette tape, HOW TO REDUCE TEST ANXIETY, reduces the high level of anxiety that is most often present in students having difficulty in math. "Teaching students how to learn mathematics, accompanied by helping them reduce their test anxiety has proven most effective," reports Dr. Nolting.

A key speaker at numerous regional and national educational conferences and conventions, Dr. Nolting has been widely acclaimed for his ability to communicate with students and faculty on the subject of improving grades.

For those interested in more information concerning Dr. Nolting's consulting services, he may be reached by writing the Academic Success Press, Inc., P. O. Box 2567, Pompano Beach, FL 33072 or calling (305) 785-2034.

Academic Success Press, Inc.
P. O. Box 2567
Pompano Beach, FL 33072

Attn: Paul D. Nolting, Ph.D.
Tele: (305) 785-2034

Our learning institution would like to consider having Dr. Paul Nolting consult with us in forming the following:

_____ General study skills courses.

_____ Mathematics study skills couse, using *WINNING AT MATH* and *HOW TO REDUCE TEST ANXIETY* as the required text/tape.

_____ Other _____

_____

_____

_____ We would like Dr. Nolting to speak at our school, college, university, club, organization, business, etc. We have enclosed details of our request.

_____ We are interested in private consultation with Dr. Nolting, and have enclosed details of our request.

Name: _____

Institution/Business: _____

Address: _____

_____

Telephone No.: _____

Best time to call: _____

OTHER WINNING BOOKS AND CASSETTE TAPES

BY DR. PAUL NOLTING

HOW TO REDUCE TEST ANXIETY is truly your key to higher test scores and better grades. No longer do you need to suffer the anxiety so commonly experienced during crucial test-taking times. Dr. Paul Nolting teaches you how to relax before and during tests to help you comfortably score higher on tests. Side one of the cassette tape covers test anxiety and short-term relaxation procedures effective in reducing it. Side two teaches the more effective long-term procedure. Proven effective. $9.95

HOW TO ACE TESTS by Dr. Paul Nolting is also your key to higher test scores and better grades. The test-taking techniques taught on this cassette have been proven effective by extensive research. No longer do you need to be penalized for being a poor test-taker, scoring far below your potential. Yes, you too can ace tests! $9.95.

WINNING AT MATH: Your Guide To Learning Mathematics Through Successful Study Skills by Paul D. Nolting, Ph.D., learning specialist. Every student must pass math courses to graduate. Doing poorly in math can decrease your career choices and even prevent you from graduating. *Winning At Math* is the book which will help you improve your math grades — quickly and easily. *Winning At Math* will help you . . .

★ Identify your personal math strengths and weaknesses, and determine how best to strengthen your weak points.

★ Understand the special study approach required for effectively learning math.

★ Understand causes of math test anxiety and learn how to relax at test time.

★ Develop time management skills for effective study.

★ Improve both listening and math note-taking skills.

★ Learn effective math textbook reading techniques.

★ Learn how to create a positive study environment.

★ Learn critical memory techniques for learning math.

★ Improve your math test-taking and test analysis skills.

★ Learn how to take control over mathematics and win!

Winning at Math is your key to success in mathematics! $12.95.

WHAT OTHERS SAY ABOUT
WINNING AT MATH

By utilizing the various techniques taught in *Winning At Math*, both my wife and I have made straight A's in our respective math courses — an unbelievable accomplishment for us! Meanwhile, our overall study skills are dramatically improved.

> Michael Mitchem
> Sarasota, FL

Winning At Math is a thorough and well conceptualized, yet practical, guide to improving math competencies. It should be recommended by every introductory level math instructor as a tool for his/her students.

> Counselor R. Craig Williams, M.A.
> College of St. Thomas
> St. Paul, Minnesota

This course (using *Winning at Math* and the cassette tape "How To Reduce Test Anxiety") has worked for me. By reducing test anxiety, I have gone from an "F" to an "A" in math.

> Hugh G. Steveley,
> Manatee Community College,
> Bradenton, Florida

We are going to use *Winning At Math* in all ten sections of our Study Habits course for math . . .

> Dr. Lewis Edwards
> Chairman, Mathematics Department
> Valencia Community College
> Orlando, Florida

Behind many failing students' apparent lack of interest and motivation, often lies a student frustrated with his lack of success in math. *Winning at Math* offers practical advice for these students. It is a valuable aid for students at every level.

> Susan R. Baker
> 'MathCounts' Sponsor
> Math Advancement Instructor
> Cimarron Middle School
> Edmond, Oklahoma

Now students of math have a resource that will help them set goals, establish good study habits, prepare for tests and reduce test anxiety. *Winning At Math* should be required reading for all high school and college students.

Jack C. Stiefel
Instructor
Center of Personalized Instruction
Vero Beach, Florida

This book is very helpful in convincing students that they can succeed in math. Its logical and systematic approach is readily accepted by students who need to understand reasons for their prior failures as well as seeing how they can succeed.

Bill Reineke
Math Lab Suppervisor
Manatee Community College
Bradenton, Florida

Winning At Math has excellent insight on the reasons students have problems in math. It gets directly to the facts students need to know. There is a definite need for this book in community colleges.

Skip Fotch
Director of Student Affairs
Matian Community College
Kentfield, CA

We are using *Winning At Math* as a required math study habits textbook in our developmental math classes.

Ray Woods
Math Department Chairman
Manatee Community College
Bradenton, Florida

ACADEMIC SUCCESS PRESS, INC.
LEARNING SPECIALISTS

LIGHT YEARS AHEAD IN LEARNING

Paul D. Nolting, Ph.D.
Learning Specialist

The "Light Years Ahead in Learning" Company, Academic Success Press, Inc., continues to move ahead in developing new ideas, improved ideas, specially developed ideas — to help students succeed.

Academic Success Press, Inc. is dedicated to making the classroom role less difficult, while producing results in students. We want to transform the classroom into a winning environment where educators and students take quantum leaps forward in unleashing hidden abilities, talents and skills through new and inventive learning techniques based on sound academic research.

The principle author and consultant to Academic Success Press, Inc. is Learning Specialist Paul D. Nolting, Ph.D. Over the past fifteen years, Dr. Nolting has helped thousands of students and professionals improve their ability to learn and obtain better grades — easily and quickly.

Dr. Nolting holds a B.S. degree in Psychology and an M.S. degree in Counseling and Human Systems from Florida State University, and a Ph.D. degree in Counselor Education from the University of South Florida. He is also licensed as a Counselor and a National Certified Vocational Evaluator. He is the learning specialist at Manatee Community College and an instructor at the University of South Florida, and a member of the Board of Directors of the Florida Developmental Education Association. His book, *WINNING AT MATH — YOUR GUIDE TO LEARNING MATH THE QUICK AND EASY WAY*, was selected Math Book of the Year by the National Association of Independent Publishers. His two audio cassettes, *HOW TO REDUCE TEST ANXIETY* and *HOW TO ACE TESTS* were also winners (in the audio cassette division) of awards in the same NAIP competition. "Dr. Nolting," says *Publisher's Report*, "is an innovative and outstanding educator and learning specialist."

As you review this catalog from Academic Success Press, Inc., please take a minute to consider each and every offer made. Any one — or all — of the educational aids noted just could make the difference in how students excel!

ACADEMIC SUCCESS PRESS, INC.
P. O. Box 2567
Pompano Beach, FL 33072
Telephone No. (305) 785-2034

LIGHT YEARS AHEAD IN LEARNING

CONTENTS

WINNING AT MATH
Your Guide To Learning Mathematics Through Successful Study Skills
by Paul D. Nolting, Ph.D.
A Study Habits Book

BEST MATH BOOK
OF
THE YEAR
National Association of
Independent Publishers

AWARD FOR EXCELLENCE
Florida Publishers Group

Winning At Math deserves space in the bookstore of every college. The student determined to overcome past difficulties will find a solid game plan for success.
Mathematics Teacher

Every student must pass math courses to graduate. Doing poorly in math can decrease your career choices and even prevent you from graduating. **WINNING AT MATH** is a book which will help you improve your math grades — quickly and easily. **WINNING AT MATH** will help you . . .

- Identify your personal math stengths and weaknesses, and determine how best to strengthen your weak points.
- Understand the special study approach required for effectively learning math.
- Understand causes of math test anxiety and learn how to relax at test time.
- Develop time management skills for effective study.
- Improve both listening and math note-taking skills.
- Learn how to take control over mathematics and win.

WINNING AT MATH is your key to success!

ISBN: 0-940287-19-6
LC: 91-19306

Trade Softcover
240 Pages / $12.95

TEACHERS' MANUAL ON **WINNING AT MATH** AVAILABLE. SEE ORDER FORM.

WINNING AT MATH STUDY AIDS

WINNING AT MATH:
Your Guide To Learning Mathematics Through Successful Study Skills
Four 90-minute Audio Cassettes — A Packaged Set
ISBN: 0-940287-20-X $29.95

The narrative of the award-winning textbook, **WINNING AT MATH**. Especially helpful to the visually impaired, learning disabled, or for those who may find reading textbooks a tiresome experience. Here's a welcome addition to any math classroom, the perfect answer to the complaint, "But I don't have time to read!" Now there will simply be "no excuse!"

WINNING AT MATH:
Study Skills Computer Evaluation Software
ISBN: 0-940287-14-5 $49.95

UP-DATED VERSION

This is a computerized math study skills evaluation with personalized printed prescription for success. The program suggests appropriate material from this catalog to improve student learning.

- The results can be seen on the screen and then printed.
- All capital letters make the program easy-to-read.
- Pencil and paper version is included; these can be used in the classroom and returned to the student.

Since math study skills must be evaluated to prescribe effective learning strategies, this method is an easy and simple approach to the problem.

Based on the proven learning principles in **WINNING AT MATH**, students can obtain a quick computerized math study skills evaluation with a personalized printed prescription for success. Likewise, this program will suggest learning and test accommodations for the physically disabled or learning disabled student. Math instructors can improve their students' success rate by requiring the computer math study skills evaluation as a homework assignment. Also, this software is an excellent supplement to the math lab or learning lab existing software.

- Available in both IBM-DOS or Apple Versions with UNLIMITED USE.
- Includes SITE LICENSE with UNLIMITED USE.
- UP-DATED VERSION

SUCCESSFUL MATH
STUDY SKILLS

**Easy-to-Use/Step-by-Step
Guide to Higher Math Grades
For
Middle and Secondary
School Students**

**by
Paul D. Nolting, Ph.D.
William A. Savage, M.S.
ISBN: 0-940287-18-8
8½ x 11 Softcover
204 pp / $12.95**

Here's a comprehensive book on math study skills specially written and designed for middle and secondary school students.

Since every student must pass math courses to graduate, **SUCCESSFUL MATH STUDY SKILLS** is the answer to this problem.

SUCCESSFUL MATH STUDY SKILLS will help the student . . .

- Identify personal math strengths and weaknesses, and determine how best to strengthen weak points.

- Understand the special study approach required for effectively learning math.

- Develop time management skills for effective study.

- Improve both listening and math note-taking skills.

- Learn how to take control over mathematics and win!

Working with a combination of the materials that Dr. Nolting produced in both student workshops and classes, we have experienced significant improvement in student performance and attitude, as well as an enhancement of teacher enthusiasm for providing new learning strategies with core curriculum.

Kevin D. Flynn
Asst. Principal
Riverview High School
Sarasota, FL

AUDIO CASSETTES

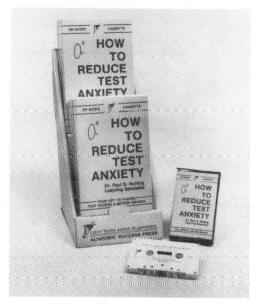

HOW TO REDUCE
TEST ANXIETY
Paul D. Nolting, Ph.D.

HOW TO REDUCE TEST ANXIETY is truly your key to higher test scores and better grades. It will teach you how to relax before and during tests and can help you improve your test scores by as much as two letter grades. No longer do you need to suffer the anxiety so commonly experienced during crucial test-taking times.

This tape has been proven effective and designed in such a fashion that it can be easily used by anyone to reduce test anxiety.

- 5 Quick & Easy-to-Learn Relaxation Techniques
- Explains the Reasons for Test Anxiety
- Long-Term Relaxation Procedure on Side 2 is Guaranteed-to-Work

You can learn these simple exercises in the comfort of your home. Let this tape assist you in acquiring a relaxed confidence when test-taking times roll around!

ISBN: 0-94027-01-3
LC: 87-70026
An Audio Cassette
$9.95

AUDIO CASSETTES

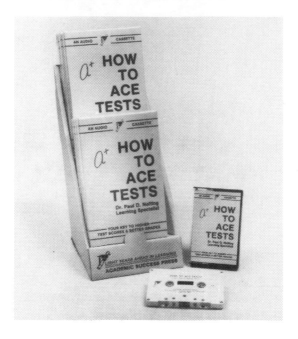

HOW TO ACE TESTS
Paul D. Nolting, Ph.D.

Can be used in conjunction with any subject taught!

HOW TO ACE TESTS is your key to higher test scores and better grades. The test-taking techniques taught on this tape have been proven effective by extensive research. No longer do you need to be penalized for being a poor test-taker, scoring far below your potential!

HOW TO ACE TESTS can be used easily by anyone wishing to increase test scores on any type of tests. It teaches you . . .

- How to predict test questions
- How to prepare for tests
- The *five* major points to becoming test-wise
- The *six* major steps to diagnosing test-taking weaknesses
- The *ten* steps to better test taking
  . . . plus much, much more.

ISBN: 0-940287-02-1
LC: 87-70027
An Audio Cassette
$9.95

STRATEGY CARDS

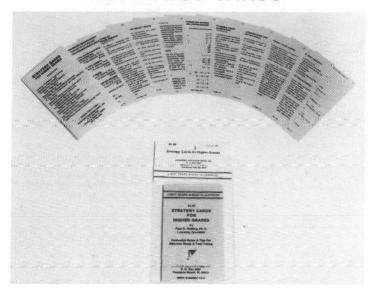

- 10 Steps To Better Test-Taking
- Reducing Test Anxiety
- 10 Steps For Reading Material — Other than Math
- Reworking Your Notes
- 10 Steps For Reading Math Material
- If You Don't Understand The Math Reading Material
- Steps For Doing Your Math Homework
- Translating English Terms Into Algebraic Symbols
- Steps For Taking An Objective Test
- Answering An Essay Test With Several Questions
- Key Words On Essay Tests
- 6 Types Of Test-Taking Errors

STRATEGY CARDS FOR HIGHER GRADES, a series of ten cards, capture invaluable rules for student study and test taking. These cards make a handy supplement to any study habits course, math course or lab course and reinforce information provided within textbooks. Transforms every course a student may take into a winner! Strategy Cards available:

A series of ten, pocket-sized (3 x 5 inch) cards
ISBN: 0-940287-10-2 $4.95

A series of five 8 ½ x 11 inch cards
for instructors and for the visually impaired
ISBN: 0-940287-11-0 $9.95

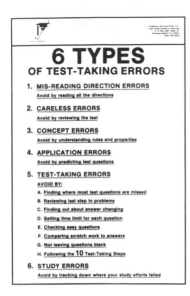

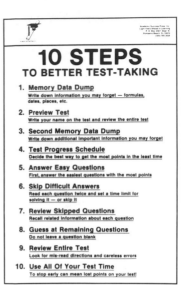

Here are invaluable classroom or lab posters, 24 by 36 inches. Produced on a bright federal yellow background — the most arresting of all colors — and printed in underlined, large black type for quick review and easy reading, no classroom or lab should be without them! Posters available:

COLLEGES/UNIVERSITIES

TEN STEPS FOR SOLVING STORY PROBLEMS   **$6.95**
TRANSLATING ENGLISH TERMS INTO
    ALGEBRAIC SYMBOLS   **$6.95**
TEN STEPS TO BETTER TEST-TAKING   **$6.95**
 SIX TYPES OF TEST-TAKING ERRORS   **$6.95**
TEN STEPS FOR DOING YOUR MATH/
    SCIENCE HOMEWORK   **$6.95**

MIDDLE SCHOOLS/HIGH SCHOOLS

TEN STEPS FOR SOLVING STORY PROBLEMS   **$6.95**
TRANSLATING ENGLISH TERMS INTO
    ALGEBRAIC SYMBOLS   **$6.95**
TEN STEPS TO BETTER TEST-TAKING   **$6.95**
 SIX TYPES OF TEST-TAKING ERRORS   **$6.95**
TEN STEPS FOR DOING YOUR MATH/
    SCIENCE HOMEWORK   **$6.95**

BOOKS

THE EFFECTS OF COUNSELING AND STUDY SKILLS TRAINING ON MATHEMATICS ACADEMIC ACHIEVEMENT is a book for educators based on Dr. Paul Nolting's dissertation research and intervention procedures that dramatically improved students' math grades. This book discusses the problems faced by today's math educators and counselors. An extensive literature review on factors that affect mathematics achievement (intelligence, study skills, anxiety, locus of control, motivation, counseling) is thoroughly discussed. The original classroom intervention design and results are explained. Overall conclusions and recommendations which you can apply to your school are included. This book is a must for math instructors and counselors who want to be up-to-date on math achievement research and need a successful blueprint for helping students succeed in math.

THE EFFECTS OF COUNSELING & STUDY SKILLS TRAINING ON MATHEMATICS ACADEMIC ACHIEVEMENT
Paul D. Nolting, Ph.D.
ISBN: 0-940287-13-7
8 ½ x 11/Trade Softcover/Second Edition
Up-dated & Revised Edition
1990 Research Included
$29.95

THE EFFECTS OF COUNSELING & STUDY SKILLS TRAINING ON MATHEMATICS has an extensive background on variables that effect math success. It gives excellent suggestions on how math instructors and counselors can help students become more successful in mathematics. Excellent support text for instructors teaching math study skills.

Dr. Robert Rapilje
Mathematics Instructor
Seminole Community College
Sanford, FL

HOW TO DEVELOP A MATH
STUDY HABITS COURSE/WORKSHOP

The math study habits course is designed to supplement students who are enrolled in a math course. The math study habits course is different from general study skills courses which have little success in helping math students. The math study habits course concentrates on the 25% of a math student's achievement based on affective characteristics — study skills, anxiety, test-taking, and motivation, while the math instructor teaches the students cognitive math skills.

Students enrolled in basic math to calculus have reported success after success by taking the math study habits course with their math course. Most students make one letter grade higher in math while taking the math study habits course and some students have gone from an "F" to an "A" in math during one semester. This team approach of teaching students math cognitive skills and study habits is a future trend in helping students.

Manatee Community College is now in its fifth year of offering the highly successful math study habits course using **WINNING AT MATH** and **HOW TO REDUCE TEST ANXIETY** as the required course materials. The book and cassette tape are wrapped together in plastic and sold for $24.95. The **STRATEGY CARDS FOR HIGHER GRADES** are optional for the course. In fact, students in developmental math courses are required to take the math study habits course along with their math course. (Other math students who are having difficulty in their course or who want to make an "A" instead of a "B" also take the course.)

Math study habits courses are generally one-hour credit, taught in a classroom or as independent study through the math lab. The course focus is on diagnosing student's affective learning charactertistics and teaching effective learning techniques to improve students' math study habits, test-taking skills, while reducing test anxiety.

Study skills instructors, counselors, or math instructors can teach math study habits. These instructors do not have to "know math" or know techniques to reduce student's test anxiety. The **TEACHER'S MANUAL** explains the purpose and teaching objectives, suggested additional topics, discussion, questions and answers to the chapter assignments for each **WINNING AT MATH** chapter. To teach the students how to reduce their test anxiety, you can play the cassette tape, **HOW TO REDUCE TEST ANXIETY**. "Doing Math" as a part of the classroom teaching is not required because the students are learning math in their math classes.

(Continued page 13)

Several strategies may be used in developing a math study habits course: (1) use an already existing course that no one is using: (2) designate certain sections of the course already being used just for math study habits; (3) use a special topics course; and (4) develop your own course. To obtain additional information on establishing a math study habits course, order the Math Study Habits Course/Workshop.

If you do not have enough time to develop a math study habits course, think about developing math anxiety/study habits workshops. You can run a mini-math workshop or a full-length math workshop. You can also add the **WINNING AT MATH** book as a supplemental text book. Students who had a "C" in their previous math course or are repeating the math course need the **WINNING AT MATH STUDY SKILLS COMPUTER PROGRAM**, too. The student should then be sent to the bookstore to purchase learning materials suggested in the program. To obtain additional information, order:

HOW TO DEVELOP YOUR OWN MATH
STUDY SKILLS COURSE/WORKSHOP

ISBN: 0-940287-16-1 $8.95

This course (using *WINNING AT MATH*) and the cassette tape (*HOW TO REDUCE TEST ANXIETY*) has worked for me. By reducing test anxiety, I have gone from an "F" to an "A" in math.

> Hugh G. Steveley
> Manatee Community College
> Bradenton, FL

MATH STUDY SKILLS SUCCESS KIT

Students, who want to improve their math grades, need their math study skills diagnosed and prescribed for successful learning strategies. Most students will need help in several areas such as test anxiety reduction, study strategies, and test-taking skills to become successful in math. If you are going to help several math students, then their prescribed treatments will be different. For these reasons the "Math Study Skills Success Kit" was developed to be a total package of diagnosis and treatment of math learning problems. The "Kit" consists of:

Winning At Math Study Skills
** Computer Evaluation Software**
Winning At Math (book)
How to Reduce Test Anxiety (audio cassette)
How to Ace Tests (audio cassette)
Strategy Cards For Higher Grades (3" X 5")

Regular price is $87.75 — Success Kit price $74.95

Working with a combination of the materials that Dr. Nolting produced in both student workshops and classes, we have experienced significant improvement in student performance and attitude, as well as an enhancement of teacher enthusiasm for providing new learning strategies within core curriculum.

Kevin D. Flynn
Assistant Principal
Riverview High School
Sarasota, FL

STRATEGY CARDS FOR HIGHER GRADES have helped my students with learning problems. Students use these cards in test situations as an auxiliary learning aid. It works.

Carol DeSouza
Counselor
University of Mass.
Boston, MA

Dear Lab Director:

Lab Directors have an opportunity to help thousands of students through one-to-one interaction and educational support materials. Unfortunately there is not enough lab staff to help all of your students. To solve this dilemma, which causes many students not to succeed, Academic Success Press, Inc. has developed proven educational material that can "act as" additional staff. Using the computerized or written version of the Math Study Skills Evaluation, **Lab Directors** can pinpoint student learning problems and suggest proven treatment procedures. The treatment can be in the form of **WINNING AT MATH, STRATEGY CARDS FOR HIGHER GRADES, CLASSROOM POSTERS, HOW TO REDUCE TEST ANXIETY**, and **HOW TO ACE TESTS**. The treatment procedures have also been successful in helping students with learning disabilities. **Lab Directors** can purchase these materials for lab use or require the bookstore to stock the material for student purchase. Either way these materials can "act as" additional staff members and improve your students' grades.

Sincerely,

Paul D. Nolting, Ph.D.
Learning Specialist

We have been very effective in diagnosing math study skills problems by using your computer program — *WINNING AT MATH* Study Skills Evaluation Computer Software. By correcting the math study skills early enough to make a difference, immediate results are obtained.

Bill Reineke
Math Lab Director
Manatee Community College
Bradenton, FL

STRATEGY CARDS FOR HIGHER GRADES are ideal to help students with learning problems. The cards are a constant reminder on how to do your homework and take tests — among the other valuable suggestions. We want our students to use them. The cards can help any student improve their learning.

WINNING AT MATH is an excellent book to teach our students how to study math. We use *WINNING AT MATH* as an individual treatment procedure to help our students do better in math.

Walter Johnson
Counselor
Valencia Community College
Orlando, FL

Dear Educator:

Let me introduce myself to you . . .

I have been working with developmental students for over ten years. As a college Learning Specialist, I have helped thousands of academically disadvantaged, physcially disabled, or learning disabled students improve their grades.

I have focused on the 25% of a student's academic achievement which is based on affective characteristics — study skills, anxiety, test-taking, and motivation. Most colleges only focus on these skills through a general study course, hoping the students can generalize or remember how to use these study skills in their other classes. However, as you know, many of our students do not possess the required skills to apply or recall study skills to their future courses. A good example of this is in mathematics. General study skills courses usually only help students in non-math/science areas. A different approach is needed to help students improve their math study skills.

With this in mind, I invested five years of research and a Ph.D. dissertation on **THE EFFECTS OF COUNSELING AND STUDY SKILLS TRAINING ON MATHEMATICS ACADEMIC ACHIEVEMENT**, where I developed materials which have proved themselves many times over in helping students pass math as well as other subjects. I developed a one-hour credit math study habits course which is extremely successful and which non-math instructors can easily teach by using The Teacher's Manual. The course materials are **WINNING AT MATH** and **HOW TO REDUCE TEST ANXIETY.**

My years of research have also indicated that our students must be reminded to use the math or general study skills. However, it is impossible for us to remind all of our students to use these study skills. One solution is to develop materials such as visual reminders (posters), auditory reminders (cassette tapes) and easily accessible reminders (Strategy Cards) which students can see or purchase as reminders of their own. This is how you can help thousands of students each year.

The posters can be put up in your classrooms or labs. The cassette tapes can be used or reserved in labs or purchased in the bookstore. The Strategy Cards can be a required or optional material for your study skills course. This is an excellent way to help your students improve their math study skills if you don't have a math study habits course.

You cannot help all the students you want unless you get help yourself. Let Academic Success Press give you that help!

Sincerely,

Paul D. Nolting, Ph.D.
Learning Specialist

WHAT OTHERS SAY

THE EFFECTS OF COUNSELING & STUDY SKILLS TRAINING ON MATHE-MATICS ACADEMIC ACHIEVEMENT gives a clear understanding of the factors that affect math achievement and that teaching students math study habits improves their math grades. It also indicates that counselors and math instructors need to work together to increase the academic success of math students. The book is a must for counselors and math instructors to obtain the research background to help students.

Dr. Nancy S. McGarry
St. Petersburg Junior College
St. Petersburg, FL

WINNING AT MATH has excellent insight on the reasons students have problems in math. It gets directly to the facts students need to know. There is a definite need for this book in community colleges.

Skip Fotch
Director of Student Affairs
Marian Community College
Kentfield, CA

Behind many failing students' apparent lack of interest and motivation, often lies a student frustrated with his lack of success in math. *SUCCESSFUL MATH STUDY SKILLS* offers practical advice for these students. It is a valuable aid for students at every level.

Susan R. Baker
'MathCounts' Sponsor
Math Advancement Instructor
Cimarron Middle School
Edmond, OK

I highly recommend the two audio cassette tapes — *HOW TO ACE TESTS* and *HOW TO REDUCE TEXT ANXIETY* — for the students who have anxiety or test-taking problems. Especially good for non-traditional college students who feel uneasy about class performance.

Norma Caltagirone
Counselor
Hillsborough Community College
Tampa, FL

ORDER FORM

BOOKS

**WINNING AT MATH: Your Guide To Learning
Mathematics Through Successful Study Skills**　　No. copies _____ X $12.95 = $ _____

SUCCESSFUL MATH STUDY SKILLS　　No. copies _____ X $12.95 = $ _____

**THE EFFECTS OF COUNSELING & STUDY
SKILLS TRAINING ON MATHEMATICS**　　No. copies _____ X $29.95 = $ _____

**HOW TO DEVELOP YOUR OWN MATH STUDY
SKILLS COURSE/WORKSHOP**　　No. copies _____ X $ 8.95 = $ _____

AUDIO CASSETTE TAPES

HOW TO REDUCE TEST ANXIETY　　No. copies _____ X $ 9.95 = $ _____

HOW TO ACE TESTS　　No. copies _____ X $ 9.95 = $ _____

**WINNING AT MATH: Your Guide To Learning
Mathematics Through Successful Study Skills**　　No. copies _____ X $29.95 = $ _____

STRATEGY CARDS

STRATEGY CARDS FOR HIGHER GRADES
3" x 5", Pocket-sized　　No. copies _____ X $4.95 = $ _____

STRATEGY CARDS FOR HIGHER GRADES
8 ½" x 11"　　No. copies _____ X $9.95 = $ _____

POSTERS

COLLEGES/UNIVERSITIES

**TEN STEPS FOR SOLVING
STORY PROBLEMS**　　No. copies _____ X $6.95 = $ _____

**TRANSLATING ENGLISH TERMS
INTO ALGEBRAIC SYMBOLS**　　No. copies _____ X $6.95 = $ _____

TEN STEPS TO BETTER TEST-TAKING　　No. copies _____ X $6.95 = $ _____

SIX TYPES OF TEST-TAKING ERRORS　　No. copies _____ X $6.95 = $ _____

**TEN STEPS FOR DOING YOUR MATH/
SCIENCE HOMEWORK**　　No. copies _____ X $6.95 = $ _____

MIDDLE SCHOOLS/HIGH SCHOOLS

**TEN STEPS FOR SOLVING
STORY PROBLEMS**　　No. copies _____ X $6.95 = $ _____

**TRANSLATING ENGLISH TERMS
INTO ALGEBRAIC SYMBOLS**　　No. copies _____ X $6.95 = $ _____

TEN STEPS TO BETTER TEST-TAKING　　No. copies _____ X $6.95 = $ _____

SIX TYPES OF TEST-TAKING ERRORS　　No. copies _____ X $6.95 = $ _____

**TEN STEPS FOR DOING YOUR MATH/
SCIENCE HOMEWORK**　　No. copies _____ X $6.95 = $ _____

COMPUTER SOFTWARE

**WINNING AT MATH: Study Skills Computer
Evaluation Software**
☐ IBM Compatible　　☐ Apple　　☐ 3 ½"　　☐ 5 ¼"　　No. copies _____ X $49.95 = $ _____

SPECIAL OFFERS

[ ] **MATH STUDY SKILLS SUCCESS KIT**　　No. copies _____ X $74.95 = $ _____

(Over)

[ ] **TEACHERS' MANUAL FOR
WINNING AT MATH**
FREE with the purchase of
20 Audio Cassettes and/or books!

No. copies _____ X $8.95 = $ _____

FREE

TOTAL PURCHASE $ _____

Sales Tax $ _____
(FL residents add 6%)
Shipping & Handling $ _____
  $3.00 First Item
  $1.50 Each Add'l. Item $ _____

YOUR TOTAL $ _____

Make all checks or money order payable to Academic Success Press, Inc.

Purchase Order No. _____ [ ] Check or Money Order Enclosed

[ ] Visa [ ] MasterCard Card No. _____ Exp. Date _____

Signature _____ Telephone No. (____) _____

Name _____ Institution _____

Mailing Address _____

MAIL ALL ORDERS TO:

**ACADEMIC SUCCESS PRESS, INC.
P.O. BOX 2567
Pompano Beach, FL 33072
(305) 785-2034**